TECHNICAL
REPORT

T0306267

Assessment of the AHRQ Patient Safety Initiative

Focus on Implementation and Dissemination Evaluation Report III (2004–2005)

Donna O. Farley, Cheryl L. Damberg,
M. Susan Ridgely, Melony E. Sorbero,
Michael D. Greenberg, Amelia Haviland,
Robin C. Meili, Stephanie S. Teleki, Lily Bradley,
Jacob W. Dembosky, Allen Fremont, Teryl K. Nuckols,
Rebecca Shaw, Stephanie L. Taylor, Hao Yu

Prepared for the Agency for Healthcare Research and Quality

RAND HEALTH

This work was prepared for the Agency for Healthcare Research and Quality under contract No. 290-02-0010. The research was conducted in RAND Health, a division of the RAND Corporation.

Library of Congress Cataloging-in-Publication Data

Assessment of the AHRQ patient safety initiative : focus on implementation and dissemination evaluation report III (2004–2005) / Donna O. Farley ... [et al.].
p. cm.
The third annual evaluation report covering the period from October 2004 through September 2005, which builds upon two previous evaluation reports covering the periods from October 2002 through September 2003 and from October 2003 through September 2004, respectively.
Includes bibliographical references.
ISBN 978-0-8330-4217-0 (pbk. : alk. paper)
1. Medical errors—Prevention—Government policy—United States. 2. Iatrogenic diseases—Prevention—Government policy—United States. 3. Patients—United States—Safety measures. I. Farley, Donna. II. Title: Assessment of the Agency for Healthcare Research and Quality patient safety initiative.
[DNLM: 1. Medical Errors—prevention & control—United States. 2. Data Collection—United States. 3. Government Programs—United States. 4. Information Dissemination—United States. 5. Outcome and Process Assessment (Health Care)—United States. 6. Program Evaluation—United States. WB 100 A838 2007]

R729.8.A872 2007
610—dc22

2007037735

Published 2007 by the RAND Corporation
1776 Main Street, P.O. Box 2138, Santa Monica, CA 90407-2138
1200 South Hayes Street, Arlington, VA 22202-5050
4570 Fifth Avenue, Suite 600, Pittsburgh, PA 15213-2665
RAND URL: http://www.rand.org/
To order RAND documents or to obtain additional information, contact
Distribution Services: Telephone: (310) 451-7002;
Fax: (310) 451-6915; Email: order@rand.org

PREFACE

In 2000, the U.S. Congress mandated the Agency for Healthcare Research and Quality (AHRQ) to take a leadership role in helping health care providers reduce medical errors and improve patient safety. AHRQ is fulfilling that mandate through a patient safety research and development initiative, which is now at the end of its fourth year of operation. In September 2002, AHRQ contracted with RAND to serve as the patient safety evaluation center for this initiative. The evaluation center is responsible for performing a longitudinal, formative evaluation of the full scope of AHRQ's patient safety initiative and for providing regular feedback to support the continuing improvement of the initiative over the four-year evaluation period.

This is the third annual evaluation report prepared by RAND, covering the period from October 2004 through September 2005. It builds upon two previous Evaluation Reports—*Context and Baseline* (Report I) and *Moving from Research to Practice* (Report II)—which cover the periods from October 2002 through September 2003, and from October 2003 through September 2004, respectively (Farley et al., 2005; Farley et al., 2007). The content and format of each report are designed to provide a stable structure for the longitudinal evaluation; the results of each year's assessment contribute to a cumulative record of the initiative's evolution.

This report provides an update on recent changes in the policy context that frames the AHRQ patient safety initiative, and it documents the evolution and current status of the priorities and activities being undertaken. The emphasis of this third report is on assessing the contributions of the health information technology projects to the patient safety initiative and on actions being taken for dissemination and broad adoption of improved patient safety practices. The report also presents baseline data on selected measures for evaluating the effects of the initiative on patient outcomes and other stakeholders. Implications of the evaluation findings are discussed with respect to future AHRQ policy, programming, and research, and suggestions are presented for strengthening AHRQ activities as the initiative continues to move forward.

The contents of this report will be of interest to national and state policymakers, health care organizations and clinical practitioners, patient-advocacy organizations, health researchers, and others with responsibilities for ensuring that patients are not harmed by the health care they receive.

This work was sponsored by the Agency for Healthcare Research and Quality, Department of Health and Human Services, for which James B. Battles, Ph.D., serves as project officer.

CONTENTS

FIGURES

TABLES

EXECUTIVE SUMMARY

As of October 2005, it has been four years since the U.S. Congress funded the Agency for Healthcare Research and Quality (AHRQ), in the Department of Health and Human Services (DHHS), to establish the national patient safety research and implementation initiative. AHRQ contracted with the RAND Corporation in September 2002 to serve as the evaluation center for this initiative.

This report—Evaluation Report III—is the third of four annual evaluation reports to be prepared by the evaluation center. It covers the period from October 2004 through September 2005. It updates information on the current status of the AHRQ patient safety initiative and examines progress in carrying out the component activities that were identified in *Evaluation Reports I* and *II* (Farley et al., 2005; Farley et al., 2007). The recommendations we offer are intended as suggestions to help guide the agency's future strategy and activities, and they are focused on actions that AHRQ is in a position to take. In some cases, we reiterate suggestions from earlier evaluation reports; in other cases, we offer new suggestions or extensions of previous ones, based on findings from the most recent evaluation analyses conducted in 2004–2005. (See the Appendix for suggestions for AHRQ action presented in the previous two reports.)

EVALUATION FRAMEWORK

In early 2000, the Institute of Medicine (IOM) published the report *To Err Is Human: Building a Safer Health System*, calling for leadership from DHHS in reducing medical errors and identifying AHRQ as the lead agency for patient safety research and practice improvements. Mandated by the U.S. Congress to lead federal patient safety improvement activities, AHRQ can provide motivation and guidance for the activities of others and, by integrating its work with that of other organizations in both public and private sectors, it can leverage finite resources and achieve synergy through collaboration.

The Evaluation Model

The overall evaluation design is based on the Context-Input-Process-Product (CIPP) model, which is a well-accepted strategy for improving systems that encompasses the full spectrum of factors involved in the operation of a program (Stufflebeam et al., 1971; Stufflebeam, Madaus, and Kellaghan, 2000). The core model components are represented in the CIPP acronym:

- ***Context evaluation*** assesses the circumstances stimulating the creation or operation of a program as a basis for defining goals and priorities and for judging the significance of outcomes.
- ***Input evaluation*** examines alternatives for goals and approaches for either guiding the choice of a strategy or assessing an existing strategy against the alternatives.
- ***Process evaluation*** assesses progress in implementing plans relative to the stated goals for future activities and outcomes.
- ***Product evaluation*** identifies consequences of the program for various stakeholders, intended or otherwise, to determine the effectiveness of and provide information for future program modifications.

A Framework for the Process Evaluation

The process evaluation is the largest and most complex component of the evaluation because many aspects of the health system are affected by AHRQ's work and that of numerous other organizations involved in patient safety. We adopted a national perspective, the goal of which was to assess the progress of the AHRQ initiative and the activities of other agencies and organizations in the context of the larger U.S. patient safety system.

We identified five system components that are essential to bringing about improved practices and a safer health care system for patients. Together, these components provide a cohesive framework for the process evaluation. They work together to bring about improved practices and a safer health care system for patients, as shown in Figure S.1. The components are (1) monitoring progress and maintaining vigilance; (2) establishing knowledge of the epidemiology of patient-safety risks and hazards; (3) developing effective practices and tools; (4) building infrastructure for effective practices; and (5) achieving broader adoption of effective practices. Our process evaluation examined progress in strengthening each of these components.

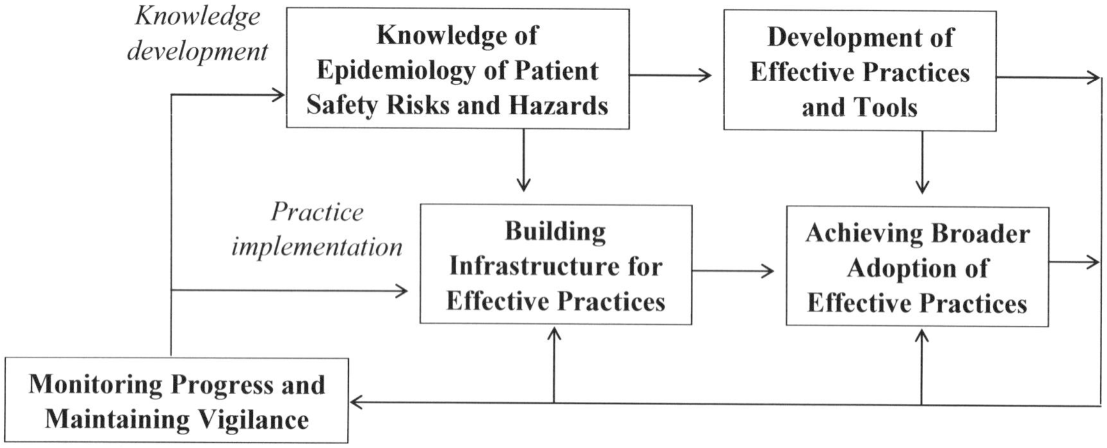

Figure S.1 The Components of an Effective Patient Safety System

The component for monitoring progress and maintaining vigilance is identified first and placed on the bottom left side of the figure, reflecting the need for early data on patient safety issues to help guide intervention choices, as well as ongoing feedback regarding progress in developing knowledge and implementing practice improvements. The top row of the figure contains the two components that contribute to knowledge development regarding patient-safety epidemiology and effective practices and tools. This knowledge is then used in the remaining two model components, which contribute to practice implementation—building infrastructure and adopting effective practices (in the second row of the figure).

FINDINGS FROM THE CONTEXT AND INPUT EVALUATIONS

Context Evaluation

The AHRQ patient safety initiative was designed within a policy context that created high expectations for achieving patient safety improvements. In *Evaluation Report I*, we identified the following implications for AHRQ, which continue to be relevant in 2005:

- *AHRQ leadership*—a clear mandate by Congress for AHRQ to provide leadership in effecting change in patient safety practices.
- *Balance between research and implementation*—the need for AHRQ to balance its traditional role of funding health services research with a shift to an implementation focus, to catalyze patient safety improvement in the health care system.
- *Resource constraints*—small appropriation of funding relative to the work at hand, including research to strengthen knowledge and actions to bring that knowledge to the health care community and increase adoption of safer practices.
- *Accountability for results*—high expectations by Congress that AHRQ demonstrates progress in improving patient safety practice and reducing harm to patients.
- *Coordination of multiple activities*—a diversity of patient safety activities undertaken by multiple public and private organizations, which requires a coordination role for AHRQ to achieve synergy among them and to encourage consistent standards of practice.

Over the past year, several new developments have raised the level of attention to patient safety, with a particular focus on using health information technology (health IT) to achieve safer care. These developments include: appropriations by Congress in September 2004 to support health IT projects ($50 million) and development of health IT standards ($10 million); the enactment of the Patient Safety and Quality Improvement Act of 2005 (Public Law 109-41) in July 2005, which the Secretary of the DHHS tasked AHRQ to implement; the development of other public- and private-sector initiatives to improve patient safety; and recent critiques of the progress of patient safety improvement.

Input Evaluation

During FY 2004, AHRQ continued the overall agency management and focus that it established in the preceding year. These include its mission and strategic plan aimed at achieving quality improvement in the health care system, work portfolios that integrate the agency's activities by topic area, and goals and objectives for the patient safety initiative.

AHRQ established the Patient Safety Research Coordinating Center (Coordinating Center) at the start of the patient safety initiative, which serves as a stimulus and facilitator of interactions among the projects funded under the initiative. A new contract was awarded in FY 2004, under which the Coordinating Center places much greater emphasis on dissemination and implementation activities and also supports broader patient safety activities within the agency. AHRQ also established a National Resource Center for Health Information Technology (Resource Center) in FY 2004, which provides technical assistance and support for the health IT grantees and assists AHRQ with managing the health IT program.

Cumulative funding for patient safety projects has generated a substantial body of work since FY 2000. The five systems-related best practices (SRBP) grants were funded in FY 2000, followed by 81 patient safety projects funded in FY 2001. The 13 patient safety challenge grants were funded in FY 2003, and 108 health IT grants and contracts were funded in FY 2004. In FY 2005, 17 grants for Partnerships in Implementing Patient Safety (PIPS) were funded, as were 16 new health IT grants and contracts.

As the patient safety initiative moved forward in FY 2004–FY 2005, AHRQ expanded its focus to include an emphasis on dissemination activities and support of implementation-oriented projects. At the same time, large field-based initiatives led by other organizations were reinforcing the dissemination of proven patient safety practices among health care providers. Within this policy and operational context, AHRQ should consider taking several categories of actions as it directs the future of the initiative:

- Engage proactively to disseminate patient safety practices.
- Collaborate actively in initiatives sponsored by others.
- Focus on establishing a national patient safety data network.
- Plan for future patient safety spending priorities.

ISSUES AND ACTION OPPORTUNITIES FROM THE PROCESS EVALUATION

Monitoring Progress and Maintaining Vigilance (Chapter 3)

AHRQ faces both opportunities and challenges in bringing about a national patient safety data network, which we have identified as a priority since the first year of the evaluation. The passage of the Patient Safety and Quality Improvement Act of 2005 (PSQIA) shifted the balance toward opportunity, because it provides for development of a network of databases as part of the larger patient safety organization (PSO) program, which offers confidentiality protections that should encourage reporting by providers. Challenges remain, however, especially in the need to reach consensus among diverse stakeholders to achieve consistency in both the measures being entered into data systems and standards for those systems. The Institute of Medicine report *Patient Safety: Achieving a New Standard of Care* (IOM, 2004) offers a useful starting point for standard development, but the responsibility for providing direction so that one set of standards can emerge probably will rest with federal agencies. AHRQ's leadership can be applied to stimulate dissemination and adoption of these standards, including close work with end users to ensure that the system designs are in fact serving their needs.

We encourage AHRQ to remain flexible in seeking ways to establish a viable national patient-safety data repository, responding to new standards or tools that arise in the field of health IT and making full use of relevant products from the earlier AHRQ-funded projects. AHRQ also should place a priority on further development of patient safety measures that could apply to aspects and settings of care for which measures do not yet exist—in particular, ambulatory care and long-term care settings.

Suggestions for AHRQ Action

- AHRQ should continue to pursue the goal of developing a national-level patient safety data capability in which multiple public and private users participate, with reinforcement by provisions in the newly enacted patient safety legislation for voluntary reporting by patient safety organizations into a network of patient safety databases.
- Using a structured consensus process involving multiple stakeholders, AHRQ should place a priority on building on the existing Patient Safety Indicators to establish a broader set of national patient safety measures that represents the most important safety aspects of the patient's health care experience in a variety of settings.

Knowledge of Epidemiology and Development of Effective Practices (Chapter 4)

The two components encompassed in the *Knowledge Development* portion of the patient safety system (Figure S.1) are addressed in one chapter, reflecting their interrelatedness as well as a shift in emphasis of the overall patient safety initiative, away from focusing on changes in knowledge of epidemiology (what are the safety issues) toward expanded development and testing of new safety practices (how to improve safety). We examine updated information on publications addressing epidemiology of patient safety risks and the contributions of AHRQ-funded projects to this information. We then assess the potential contributions to knowledge of the three sets of new health IT grants that AHRQ funded in FY 2004, totaling 104 projects, as part of our ongoing assessment of new patient safety projects as AHRQ funds them.

The existing patient safety grants have been producing a fast-growing literature on patient safety epidemiology, which can help decisionmakers assess future priorities. At the same time, the health IT projects, as with AHRQ's other patient safety projects, are addressing a diversity of patient safety issues and have the potential to contribute new knowledge and additional scientific evidence for new patient safety practices. According to reports from the health IT grantees, the funding that AHRQ has provided to them is making a significant contribution to the diffusion and adoption of health IT across a range of health-care practice settings. The partnerships across communities developed in these projects, and lessons learned by the grantees in working with them, can serve as guidance for others in enhancing their adoption of health IT.

Not surprisingly, the early experiences of the health IT projects revealed many of the same design and implementation issues identified by previous patient safety projects. They also raised numerous other issues related specifically to use of IT systems. This information is valuable for other organizations embarking on similar projects. It also can help the Resource Center to fine-tune its technical support for the grantees.

Suggestions for AHRQ Action

- AHRQ should maintain an ongoing monitoring process that uses data from the national network of patient safety databases, as well as published research, to examine shifts in trends for patient safety epidemiology in specific aspects of health care and to identify emerging safety issues that need to be addressed to ensure the safety of health care practices.
- AHRQ and the Resource Center should establish a structured program of start-up support and training to first-time grantees to help them understand their responsibilities and AHRQ's expectations.
- AHRQ should encourage and support work on development of flexible, inexpensive, IT solutions that are accessible for low-resource and rural organizations.
- AHRQ should explore mechanisms to strengthen the evaluation component of the health IT implementation projects, such as providing more evaluation technical assistance, requiring that grantees partner with evaluation researchers; or having the evaluation component be conducted independently from the health IT implementation.
- AHRQ should work with the Resource Center to clarify to grantees the functions of the Resource Center and the type of technical assistance it can provide to grantees, and to tailor technical assistance on evaluation measures and methodologies to the unique features of local projects.
- In partnership with other relevant federal agencies, AHRQ should develop clear federal guidance on standards and other requirements for interoperability of health IT, making the investment in health IT more compelling and easier for low-resource organizations.

Building Infrastructure for Effective Practices (Chapter 5)

Our evaluation identified patient safety culture, information systems, adverse-event-reporting systems, interdisciplinary teams, multi-institutional collaborations, and quality-improvement systems and measures as key components for a patient safety infrastructure. Our assessment of ARHQ's contribution to each of these components has found that, thus far, ARHQ has made its strongest contributions to building patient safety culture, information systems, adverse-event-reporting systems, and multi-institutional collaborations. It has done so through the types of patient safety projects it has funded, its partnership outreach with other organizations, its specific initiative for high-reliability organizations, dissemination of the Hospital Survey on Patient Safety Culture, the new Accelerating Change and Transforming Organizations and Networks (ACTION) and Developing Evidence to Inform Decisions About Effectiveness (DEcIDE) networks, and other network programs. In addition, our analyses have found that the Patient Safety Improvement Corps (PSIC) has stimulated actions by its participants to implement patient safety improvements in their respective organizations.

In considering what more needs to be done to ensure the establishment of effective infrastructures for patient safety practices in the health care system, it will be important to assess not only AHRQ's contributions but also those of other organizations that have been undertaking similar actions at the national, regional, and local levels. New work currently under way by AHRQ includes development of products for health care providers to strengthen interdisciplinary teams, which it plans to introduce in FY 2006, and work to strengthen quality-improvement systems and measures. Continued work also is needed for already-established programs.

The Hospital Survey on Patient Safety Culture, if used widely by hospitals across the country, could have far-reaching effects on performance. We note that hospitals (or other organizations) using the culture survey need to use the information it generates to guide actions for changing practices and attitudes, to ultimately achieve a strong safety culture. It is critical for AHRQ to capture information on the use of the survey and resulting improvement activities by hospitals, so that the effect of the survey can be documented and assessed.

The first two years of the PSIC have demonstrated the value of bringing together groups of diverse stakeholders and of providing them with intense patient safety training. With completion of the training for a third group of participants in May 2006, the original goal and scope of the PSIC will have been achieved. However, the very success of this PSIC scope of work raises questions about how to establish a sustainable infrastructure for building on the expertise and partnerships it has fostered thus far. AHRQ and the Department of Veterans' Affairs (VA) face some important decisions regarding how to approach the next generation of PSIC activities, including options for a train-the-trainer program, continued support of patient safety activities by PSIC graduates, and strategies to reach health care decisionmakers who ultimately determine how far these activities can proceed.

In considering future options for building infrastructure, AHRQ will need to choose strategically where to invest its limited resources. Engaging in partnerships is a useful strategy to leverage resources, but AHRQ will need to decide which types of partnerships might be the most fruitful to lead, and which are better led by other organizations. The new AHRQ network initiatives for high-reliability organizations, ACTION, and DEcIDE offer rich potential to contribute to the patient safety infrastructure. We encourage AHRQ to monitor these programs regularly, seeking feedback from participants, and gathering information to learn from experience and modify the programs as needed.

Suggestions for AHRQ Action

- As AHRQ continues to work with hospitals to support their use of the Hospital Survey on Patient Safety Culture, it should establish a structured monitoring process that collects data on trends in hospital use of the survey, issues identified by hospitals from survey data, and actions hospitals take to respond to those issues.
- Building upon the successful PSIC training that has reached the front-line hospital and state-level staff, AHRQ should work with relevant organizations to stimulate outreach to increase commitment to patient safety by key decisionmakers who are needed to make patient safety improvements happen.
- AHRQ should assist PSIC graduates in finding support for their continued engagement in patient safety issues, updating their skills and knowledge, and encouraging synergy among the PSIC graduates and with others in the field.
- If AHRQ and the VA pursue a PSIC train-the-trainer program, it should be a working partnership among AHRQ, the VA, and the trainers to ensure that the persons who become trainers have the needed teaching skills to fulfill the defined trainer role.

Achieving Broader Adoption of Effective Practices (Chapter 6)

To guide our assessment of AHRQ's progress in dissemination of new safe practices and tools, we developed a framework that identifies AHRQ as a change agency—an entity that works to disseminate new practices (innovations) for adoption in the field. This framework also identifies four phases of action by change agencies: problem recognition and assessment, development of usable innovations, packaging of innovations, and dissemination and diffusion of innovations (see Figure 6.1 and Table 6.1). We found that AHRQ has done extensive work in the problem recognition and assessment phase of the dissemination process. However, its activities for the remaining phases thus far have occurred on a relatively small scale. Although its dissemination activities have grown (e.g., packaging and dissemination of the AHRQ patient safety culture survey and packaging products from the Partnerships in Implementing Patient Safety grants), AHRQ has made slow progress in both the development and packaging of innovations and the dissemination of those innovations. AHRQ has yet to implement a comprehensive strategy for dissemination of patient safety practices and tools, owing to several factors, including difficulties in getting complete information from grantees on their project products and findings, and it has a tendency to emphasize communications rather than product packaging and dissemination for end users. In addition, end users have not yet had much involvement in AHRQ's process for establishing priorities for products and tools.

At the same time, initiatives for implementing proven patient safety practices have been mobilizing in the field with increasing speed. Recent national initiatives, which are attracting participation by hundreds of hospitals across the country, show that commercialization of proven patient safety products is happening in the field, which may be the best place for that work to be done. These initiatives include the Surgical Care Improvement Project (SCIP) led by the Centers for Medicare and Medicaid Services (CMS), the 100,000 Lives (now the 5 Million Lives) Campaign led by the Institute for Healthcare Improvement (IHI), and hospital transformation work by the Quality Improvement Organizations (QIOs). AHRQ will need to be able to move quickly to match the pace of these initiatives and serve as a useful resource to them.

Given this context of strong and growing activities by end users, AHRQ should focus its efforts where it offers unique capability to make the best contributions to the dissemination of patient safety practices. Although AHRQ is viewed by end users as the leader in patient safety research and knowledge, the agency is not an organization on the front line of health care delivery—which is where changes in practices need to occur to improve safety. We suggest that foremost among the unique contributions AHRQ can make will be the new knowledge and products from its patient safety projects, including establishment of priorities for patient safety actions based on syntheses of this information. A complementary contribution is service by AHRQ as a national clearinghouse for patient safety information, including results from AHRQ-funded research, products and tools for safe practices, and linkages to information provided by field-based organizations and initiatives.

Suggestions for AHRQ Action

- AHRQ should develop and implement a strategic plan that defines its focus on and roles for disseminating new patient-safety knowledge and products, with support for this work coming from an appropriate internal infrastructure and budget.
- As AHRQ continues to partner with health care systems and other implementers on dissemination activities, it should place a priority on synthesizing information and packaging products and tools from the patient safety grantees so that this information can be available to the field in a usable and timely manner.
- Building on its strength as a national information resource on the scientific basis of patient safety issues and practices, AHRQ should extend its work to further establish the Patient Safety Net (PSNet) as an integrated clearinghouse on patient safety, including linkages to information provided by other organizations, that is the "go-to" place for users across the country.

PRODUCT EVALUATION: PATIENT SAFETY OUTCOMES

Analysis of baseline trends in outcomes is a necessary first step in exploring the effect of AHRQ's patient safety activities, which is the work of the product evaluation (see CIPP definition above). We have defined the baseline period to be the late 1990s through 2003, which immediately precedes the time during which actions undertaken through the AHRQ patient safety initiative should begin to influence patient safety outcomes. Data for more-recent years will begin to become available by 2006, and these data can be used to assess whether early effects of patient safety activities are observable in trends for the patient-outcome measures.

In preparation for this assessment, exploratory analysis of patient-safety-outcome measures was performed using data from other organizations and our own analyses of encounter data. These analyses served three purposes: (1) provide information for health care providers, policymakers, standard-setting organizations, and other stakeholders regarding historical performance for selected safety outcome measures; (2) identify measurement and methodological issues that needed to be addressed to ensure the validity of the trend data used to assess effects; and (3) serve as the basis for estimating linear baseline trend lines to which future values for the measures can be compared.

Several highlights emerged from the analyses of trends in reported events and rate measures (i.e., with denominators), which will be considered as the product evaluation moves forward. Perhaps the most obvious one is the large differences in the trends observed for the reported events and measures that are rates. Occurrences of reported events have increased over time for both the Joint Commission sentinel events and the MedMARx adverse medication events. By contrast, trends in rate measures generally have changed more gradually, some moving up and others moving down, and rates for some measures have not changed at all.

For each type of measure, we identified specific issues that will affect their usability for national-level monitoring. For reported-event measures, it will be important to isolate independent effects on observed rates by underreporting of events, increased reporting as patient safety vigilance increases, and subsequent reductions in reported events as safety practices improve. For measures based on rates, trends vary across measures and over time, so it will be essential to select carefully the measures used to monitor changes in outcome rates to ensure that, as a group, they fulfill the function of serving as proxies for larger performance trends across a health care setting. In addition, it will be important to use multiple years to establish baseline trend data, which brings with it the challenge of adjusting estimates for changes in definitions for diagnosis codes, as well as in methods for calculating the measures using them.

Our examination of baseline trends in patient safety outcomes identified some important opportunities for future action by AHRQ. Ambulatory care settings are perhaps the highest priority for a development effort, but even currently existing hospital claims-based measures and datasets can benefit from additional refinement, expansion, and validation. In year four, we will expand on the work presented here by exploring the initial effects of AHRQ's activities on patient safety outcomes. We also will examine some additional sources of patient-safety process and outcomes data, and will make further efforts to contribute to the development of patient safety measures across a range of health care settings and procedures.

Suggestions for AHRQ Action

- AHRQ should harmonize and validate the capture of claims information in existing inpatient claims databases (in collaboration with CMS and other organizations), by evaluating the consistency of claims-coding practices across hospitals and regions, and by adding to the MCBS claims files the data needed to estimate patient safety measures.
- AHRQ should place a priority on developing a set of patient safety measures for ambulatory care settings, and it should foster the establishment of a data infrastructure that can support measurement for ambulatory-care patient-safety issues.
- AHRQ should work collaboratively to establish an infrastructure and procedures for regular collection of data on the use of effective patient-safety tools and practices by health care organizations, for inclusion in the national network of databases, along with reports from the organizations about the effects of those tools and practices on care processes and clinical outcomes.

NEXT STEPS

The evaluation results presented in *Evaluation Report III* have focused on the process and product evaluations performed in 2004–2005. This phase of the evaluation focused on the initiative's contributions to each of the five system components (Figure S.1), including assessment of potential contributions of the health IT projects to patient safety knowledge and practices and AHRQ's progress in activities to disseminate proven patient safety practices for broad adoption by health care providers. We also developed baseline trends for selected measures to assess effects of the initiative on patient outcomes and other stakeholders. In addition, the activities of field-based initiatives (e.g., 5 Million Lives Campaign; see Chapter 6) have become subjects for the evaluation, because they are important vehicles for the diffusion of safe practices across the health care system, and AHRQ is partnering in them as part of its dissemination strategy.

The progress of AHRQ's patient safety initiative in the five years since the publication of the IOM report *To Err Is Human* (IOM, 2000) can be summarized with respect to each of the five system components on which we organized our process evaluation. The greatest progress has been in the contributions made to development of knowledge of patient safety epidemiology and effective practices, through the patient safety projects it has funded each year, as well as to development or strengthening of infrastructure to support adoption of safety practices. The components for which progress has been slower are the establishment of a monitoring and vigilance capability and the dissemination of knowledge and products into the field for use by health care providers and other end users.

The activity mix of the patient safety initiative has been shifting, as AHRQ has increased its emphasis on the synthesis of knowledge and packaging of products emerging from its funded projects, along with packaging and dissemination of products and tools for adoption by health care providers. With small funding relative to the size of the national patient safety problem, however, the potential for AHRQ to have an effect on creating a safer health care system may be limited. Recognizing its funding constraints, AHRQ has been using a variety of approaches to leverage resources, such as cost sharing on implementation projects and working through partners.

Results from the FY 2000–FY 2001 projects, which are entering the published literature rapidly, will need to be synthesized to provide the information base for the dissemination and adoption strategies. Other AHRQ activities—such as the PSIC and the hospital culture survey—have come to maturity and are beginning to build infrastructure and influence patient safety practices across the country. Its most recently funded grant projects are addressing health IT and implementing patient safety practices, thereby continuing to build knowledge and evidence to support practice improvements.

Achieving the goal of a safer health care system will depend on the commitment of thousands of organizations, with leadership and support from agencies such as AHRQ and other government agencies. The momentum of widespread participation by providers in field-based initiatives has become highly visible this year, which bodes well for diffusion of safer practices. The enactment of the PSQIA has created another important mechanism that makes it safer for providers to report adverse events and work to prevent them, and it provides a much-needed stimulus for establishment of a national patient safety data repository.

From our observations of AHRQ's patient safety strategy, and in view of the current activities of its grantees and field organizations, we have identified several priorities that we encourage ARHQ to pursue in the near future:

- Facilitate movement toward a *national network of patient safety databases* by using the provisions in the Patient Safety and Quality Improvement Act of 2005 to encourage use of consistent data standards, and establish a set of national patient safety measures for assessing performance.
- *Identify key patient safety practices and products* from the results of the FY 2000–FY 2001 patient safety projects by synthesizing results from groups of projects addressing similar issues or practices.

- *Package and disseminate* patient safety products and tools that derive from the synthesis of project results, including development of "off-the-shelf" products that can be used readily by health care organizations.
- Update the *patient safety evidence report* to incorporate recently published results from the patient safety projects, applying standards of evidence that ensure rigorous assessment of study designs for testing patient safety practices that cannot be tested effectively using randomized control study designs.
- Assess both the role of *health IT* in achieving safer health care practices and its interface with the human aspects of care delivery, using results of the newly funded health IT grants as well as knowledge generated by other patient safety projects that have addressed use of technology for patient safety practices.
- Continue to *engage actively in field-based partnerships* that enable ARHQ to optimize its effect in stimulating broad adoption of proven patient safety practices by health care providers, within the constraints of its finite resources.

In 2005–2006, as the patient safety evaluation center embarks on its last year, the RAND team will continue gathering information on the evolution of the patient safety initiative, and we will consolidate results across the full evaluation to present cumulative information on the progress and effects of the patient safety initiative. We also will attempt to assess the experiences and lessons from the field-based patient safety initiatives discussed in this report, which we view as being central to successful diffusion of safe practices across providers. For the product evaluation, we will move into assessment of the effects of the initiative as far as possible in this limited time frame.

ACKNOWLEDGMENTS

We gratefully acknowledge the participation of numerous individuals in the evaluation process. At the national level, AHRQ staff and staff of other federal agencies and private-sector organizations involved in patient safety activities have provided useful perspectives and information on the initiative's approach and activities.

The principal investigators of the AHRQ-funded patient safety and other related projects or initiatives have also contributed valuable information through their participation in interviews and focus groups, and by providing written materials about activities relevant to the patient safety initiative. Grantees have shared their experiences in the execution of their research activities, as well as in the cross-grantee collaborative activities supported by AHRQ and its contractors. Individuals in other organizations involved in patient safety activities have also been generous with their time and information, enabling us to gain a comprehensive understanding of the growing volume of patient safety activities occurring in the field and of AHRQ's contribution to them.

Our AHRQ project officer, James Battles, has been instrumental in guiding the conceptual formation and execution of the evaluation. His support derives from a commitment to objective, formative evaluation, and to creating opportunities for learning over time, both of which provide a strong foundation for this evaluation. We also thank our RAND colleagues, Chau Pham, Susan Lovejoy, Liisa Hiatt, Scott Ashwood, and Stacy Fitzsimmons for their indispensable contributions to our data-collection and analysis processes. We acknowledge gratefully the editing done by Donna Keyser, in which she tightened up and shortened the much longer, original draft report. Finally, we thank Michal Tamuz and Marla Haims for their comments on an earlier draft of this report. Any errors of fact or interpretation are, of course, the responsibility of the authors.

ACRONYMS

ACTION	Accelerating Change and Transforming Organizations and Networks
AHA	American Hospital Association
AHIC	American Health Information Community
AHRQ	Agency for Healthcare Research and Quality
CAHPS	Consumer Assessment of Health Plans
CDC	Centers for Disease Control and Prevention
CDOM	Center for Delivery, Organization, and Markets
CEO	chief executive officer
CERT	Centers for Education and Research on Therapeutics
CIPP	Context-Input-Process-Product
CLIPS	Clinical Informatics to Promote Patient Safety
CMS	Center for Medicare and Medicaid Services
CPOE	computerized provider order entry
CP3	Center for Primary Care, Prevention, and Clinical Partnerships
CQuIPS	Center for Quality Improvement and Patient Safety
DEcIDE	Developing Evidence to Inform Decisions About Effectiveness
DHHS	Department of Health and Human Services
DoD	Department of Defense
DOQ-IT	Doctors' Office Quality–Information Technology
DRG	Diagnosis Related Group
EHR	electronic health record
EPC	Evidence-based Practice Center
FDA	Food and Drug Administration
HCUP	Healthcare Cost and Utilization Project
HIMSS	Healthcare Information Management Systems and Society
HRSA	Health Resources and Services Administration
ICD-9	International Classification of Diseases, Ninth Revision
IDSRN	Integrated Delivery System Research Network
IHI	Institute for Healthcare Improvement
IOM	Institute of Medicine
IT	information technology
KT	Knowledge Transfer
MCBS	Medicare Current Beneficiary Survey
MDS	minimum data set
MEPS	Medical Expenditure Panel Survey
MPSMS	Medicare Patient Safety Monitoring System
MSA	metropolitan statistical area
NCPS	National Center for Patient Safety
NHII	national health information infrastructure

NHQR	National Health Quality Report
NIS	national inpatient sample
NNIS	National Nosocomial Infections Surveillance System
NORC	National Opinion Research Center
NPSF	National Patient Safety Foundation
NQF	National Quality Forum
NS	not significant
OCKT	Office of Communications and Knowledge Transfer
ONC	Office of the National Coordinator of Health Information Technology
PBRN	Practice-Based Research Network
PFQ	Partnership for Quality
PI	principal investigator
PIPS	Partnerships in Implementing Patient Safety
PSET	Patient Safety Event Taxonomy
PSI	patient safety indicator
PSIC	Patient Safety Improvement Corps
PSNet	Patient Safety Net
PSO	patient safety organizations
PSQIA	Patient Safety and Quality Improvement Act of 2005
PSRCC	Patient Safety Research Coordinating Center
PSTF	Patient Safety Task Force
QAPI	quality assessment and performance improvement
QIO	quality improvement organization
QuIC	Quality Interagency Coordination Task Force
RCA	Root cause analysis
RFA	Request for Application
SCIP	Surgical Care Improvement Project
SRBP	systems-related best practices
TRIP	Translating Research Into Practice
UCSF	University of California, San Francisco
ULP	User Liaison Program
USDA	U.S. Department of Agriculture
UT–MO	Utah–Missouri
VA	Department of Veterans' Affairs

CHAPTER 1. INTRODUCTION

As of October 2005, it has been four years since the U.S. Congress funded the Agency for Healthcare Research and Quality (AHRQ) to establish the national patient safety research and implementation initiative. With these funds, AHRQ has committed to improving patient safety in the U.S. health care system by developing a comprehensive strategy for supporting expansion of knowledge about patient safety epidemiology and effective practices and by identifying and disseminating the most effective practices. AHRQ contracted with RAND Corporation in September 2002 to serve as the evaluation center for its patient safety initiative. The evaluation center is responsible for performing a longitudinal evaluation of the full scope of AHRQ's patient safety activities and for providing regular feedback to support the continuing improvement of this initiative. This report—*Evaluation Report III*—is the third of four annual evaluation reports to be prepared by the evaluation center. The report covers the period October 2004 through September 2005.

THE POLICY CONTEXT

In early 2000, the Institute of Medicine (IOM) published the report entitled *To Err Is Human: Building a Safer Health System*, calling for leadership from the U.S. Department of Health and Human Services (DHHS) in reducing medical errors, and identifying AHRQ as the lead agency for patient safety research and practice improvement (IOM, 2000). In response to the IOM report, the Quality Interagency Coordination Task Force (QuIC) identified more than 100 actions designed to create a national focus on reducing errors, strengthen the patient-safety knowledge base, ensure accountability for safe health care delivery, and implement patient safety practices (QuIC, 2000).

When the U.S. Congress established patient safety as a national priority and gave AHRQ the mandate to lead federal patient-safety-improvement activities, it provided AHRQ with funding to support related research and implementation activities. The AHRQ patient safety work is one of numerous and important patient safety initiatives being undertaken by a variety of organizations across the country. AHRQ's leadership can provide motivation and guidance for the activities of others, and by integrating its work with that of public and private organizations, the agency can leverage finite resources and achieve synergy through collaboration.

EVALUATING THE PATIENT SAFETY INITIATIVE

The CIPP Evaluation Model

Through this longitudinal evaluation, lessons from the current experiences of AHRQ and its funded projects can be used to strengthen subsequent program activities. As specified by AHRQ in the evaluation contract, the overall study design is based on the Context-Input-Process-Product (CIPP) evaluation model, which is a well-accepted strategy for improving systems that encompasses the full spectrum of factors involved in the development, operation, and outcomes of a program (Stufflebeam et al., 1971; Stufflebeam, Madaus, and Kellaghan, 2000). The core model components are represented in the CIPP acronym:

- ***Context evaluation*** assesses the circumstances stimulating the creation or operation of a program as a basis for defining goals and priorities and for judging the significance of outcomes.

- ***Input evaluation*** examines alternatives for goals and approaches for either guiding the choice of a strategy or assessing an existing strategy against the alternatives, including congressional priorities and mandates, as well as agency goals and strategies. Stakeholders also are identified, and their perspectives on the patient safety initiative are assessed.
- ***Process evaluation*** assesses progress in implementation of plans relative to the stated goals for future activities and outcomes.
- ***Product evaluation*** identifies consequences of the program, intended or otherwise, for various stakeholders to determine effectiveness and provide information for future program modifications.

Table 1.1 illustrates the sequence of the four stages of the CIPP model as applied to this program evaluation. The activities covered in this third evaluation report are shown in the shaded column. They include updates on the context and input evaluations, and continued assessment of patient-safety-initiative activities through the process evaluation. The product evaluation consists of an initial assessment of baseline data related to outcomes of the initiative, which will serve as the basis of the product-evaluation results, to be presented in the fourth and final report.

Major Stakeholder Groups Addressed

We have identified the following major stakeholder groups for the patient safety initiative, for which effects should be assessed:

- *Patients* – who receive health care services and bear the brunt of adverse health care events, have a direct stake in the occurrence of those events.
- *Providers* – including physicians, nurses, and the organizations that employ them, also have a stake in the occurrence of adverse events, as well as in the adoption of clinical and organizational practices designed to promote safety.
- *States* – that license health care providers and (in many instances) operate adverse-event-reporting systems, have a stake in tracking adverse events and in promoting remediation efforts by providers.
- *Organizations working in patient safety* – organizations that are working to promote best practices, education, and technology adoption in patient safety, and that have a stake in building collaborations to achieve those ends.
- *Federal government* – agencies in the federal government involved in patient safety activities—in particular, AHRQ and other DHHS agencies.

A Framework for the Process Evaluation

For AHRQ's patient safety initiative, the process evaluation is the largest and most complex component of the evaluation because many aspects of the health system are affected by AHRQ's work and that of numerous other organizations involved in patient safety. We identified five system components that are essential to bringing about improved practices and a safer health care system for patients, which together provide a cohesive framework for the process evaluation, as shown in Figure 1.1. Our process evaluation examined progress in strengthening each of these five system components, addressing for each component the three questions identified above: (1) Is the initiative reaching the target population(s)? (2) Are delivery and support functions consistent with program design? and (3) Are positive changes occurring as a result of these activities?

Table 1.1
Time Line for Reporting Results from the Longitudinal Evaluation of the National Patient Safety Initiative

	Contents and Time Periods of Evaluation Reports			
	Report 1: History-Sept 2003	Report 2: Oct 2003-Sept 2004	Report 3: Oct 2004-Sept 2005	Report 4: Oct 2005-Sept 2006
Context Evaluation				
Initial assessment of context	X			
Updates on context changes		X	X	X
Input Evaluation				
Assessment of goals and strategy established for the initiative	X			
Updates on changes in goals or strategy		X	X	X
Process Evaluation				
Baseline documentation of patient safety activities related to the initiative	X			
Assessment of contributions by AHRQ-funded patient safety projects to *patient safety knowledge* and *patient safety practices*	X	X	X	X
Assessment of other mechanisms used by AHRQ to strengthen patient safety practices		X	X	X
Assessment of dissemination of new knowledge to stakeholders in the field		X	X	X
Assessment of progress in adoption of effective patient safety practices		X	X	X
Product Evaluation				
Initial identification of potential outcome measures and data sources		X		
Development of data sources when feasible			X	X
Documentation of baseline trends for selected measures			X	X
Assessment of impacts of the patient safety initiative on selected measures				X
Establishment of infrastructure for AHRQ to continue and expand monitoring impacts			X	X

This system framework can represent the components of an effective system at either the national level or a more local level. At the national level, AHRQ is engaged in all of these system components, as are numerous other key organizations. The system components are defined as follows:

Monitoring Progress and Maintaining Vigilance. Establishment and monitoring of measures to assess performance improvement progress for key patient safety processes or outcomes, while maintaining continued vigilance to ensure timely detection and response to issues that represent patient safety risks and hazards.

Knowledge of Epidemiology of Patient Safety Risks and Hazards. Identification of medical errors and causes of patient injury in health care delivery, with a focus on populations that are vulnerable because they are compromised in their ability to function as engaged patients during health care delivery.

Development of Effective Practices and Tools. Development and field testing of patient safety practices to identify those that are effective, appropriate, and feasible for health care organizations to implement, taking into account the level of evidence needed to assess patient safety practices.

Building Infrastructure for Effective Practices. Establishment of the health care structural and environmental elements needed for successful implementation of effective patient safety practices, including an organization's commitment and readiness to improve patient safety (e.g., culture, information systems), hazards to safety created by the organization's structure (e.g., physical configurations, procedural requirements), and effects of the macro-environment on the organization's ability to act (e.g., legal and payment issues).

Achieving Broader Adoption of Effective Practices. The adoption, implementation, and institutionalization of improved patient safety practices to achieve sustainable improvement in patient safety performance across the health care system.

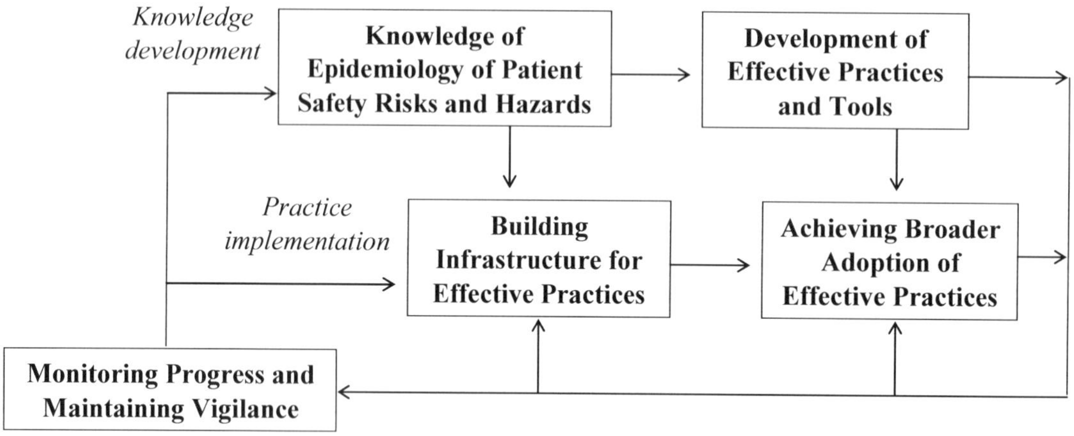

Figure 1.1 The Components of an Effective Patient Safety System

The component for monitoring progress and maintaining vigilance is identified first and placed on the left side of the figure, reflecting the need for early data on patient safety issues to help guide intervention choices. This function then continues to provide routine feedback regarding progress in developing knowledge and implementing practice improvements. The top row of the figure contains the two components that contribute to knowledge development regarding patient safety epidemiology and effective practices and tools. This knowledge is then used in the remaining two model components (in the second row of the figure) that contribute to practice implementation—building infrastructure and achieving adoption of effective practices.

Approach and Methods

The study design allows for both a national-level evaluation of the overall AHRQ patient safety initiative and a local-level evaluation of the contributions of the patient safety projects funded by AHRQ. At the national level, AHRQ is building a coordinated initiative from which the collective activities and knowledge generated across the country can be applied to improve patient safety epidemiology and practices. At the local level, our evaluation focuses on the work of the AHRQ-funded projects, which are working at various local and regional levels to generate new knowledge on patient safety epidemiology, develop new practices for preventing errors and adverse events, or test new practices or infrastructures to support practices under field conditions. The Patient Safety Research Coordinating Center (hereafter called the Coordinating Center) is funded by AHRQ to serve as a facilitator of interactions among the patient safety grantees, and to provide technical support to the grantees and AHRQ.

Numerous data-collection methods are used for the evaluation and are tailored to each specific aspect of the patient safety initiative being addressed. We use existing information from written reports and documents, Web sites, and proposals written for the patient safety projects that were awarded AHRQ funding. We also conduct open-ended interviews with numerous individuals, including AHRQ personnel, grantees, and external stakeholders, to gather information on the dynamics and issues relevant to the patient safety initiative.

ABOUT THIS REPORT

This evaluation report updates information on the current status of the AHRQ patient safety initiative and examines progress in carrying out the component activities that were identified in *Evaluation Reports I* and *II* (Farley et al., 2005; Farley et al., 2007). The recommendations we offer focus on actions that AHRQ is in a position to take and are intended as suggestions to help guide the agency's future strategy and activities. In some cases, we reiterate recommendations offered in *Evaluation Reports I* and *II*; in other cases, we offer new recommendations or expansions of previous ones based on our most recent findings. (See the Appendix for suggestions for AHRQ action presented in the previous two reports.)

The remaining seven chapters of this report are organized according to the context, input, process, and product components of the CIPP evaluation model. Chapter 2 focuses on the context and input components of the evaluation, summarizing the history leading up to funding of the patient safety initiative and presenting updated information on AHRQ's patient safety strategy, activities, and budget. Chapters 3 through 6 present assessments from our process evaluation on the progress and current status of the AHRQ patient safety initiative, which are organized according to the five-component patient safety system structure presented in Figure 1.1 and defined above. Chapter 3 addresses monitoring and vigilance, Chapter 4 addresses the two components of developing knowledge on patient safety epidemiology and practices, Chapter 5 addresses infrastructure, and Chapter 6 addresses activities for adoption of effective practices. Chapter 7 addresses the product evaluation component of the CIPP model and presents baseline trends for selected patient safety outcome measures. Chapter 8 concludes with a summary of the current status of the AHRQ patient safety initiative, and it describes the steps to be completed in the fourth and final year of this evaluation.

Unless stated otherwise, the information presented in this report is current as of September 2005. Assessment of the additional activities related to AHRQ's national patient safety initiative undertaken since that time will be included in *Evaluation Report IV*.

CHAPTER 2. CONTEXT AND INPUT EVALUATIONS

This chapter updates the information presented in *Evaluation Reports I* and *II* regarding the policy context that frames the AHRQ patient safety initiative (context evaluation), as well as the priorities and activities being pursued by AHRQ as it implements the initiative (input evaluation).

THE POLICY CONTEXT

The historical context that led to formation and funding of the AHRQ patient safety initiative may be summarized as follows:

- The science of patient safety was relatively immature as this initiative began, including limited knowledge of the epidemiology of safety in health care, an inadequate body of published research to establish evidence regarding the effectiveness of practices to improve patient safety, and lack of recognition or acceptance within the health care system that there was a "patient safety problem."
- Strong public sentiment and support for reducing health care harm to patients was stimulated by the IOM report *To Error Is Human: Building a Safer Health System*, resulting in action by Congress to make patient safety a national policy priority.
- Following a difficult period in which the agency had received criticism and had been at risk of discontinuation, under new leadership, AHRQ received re-authorization in 1999 with a new mandate from Congress, including a leadership role in patient safety. At the same time, the Quality Interagency Coordination (QuIC) Task Force also began to address patient safety, coordinating the work of its member agencies.[1]
- Congress enacted the initial appropriation of $50 million for FY 2001 and designated AHRQ to lead the federal patient safety initiative and fund needed research.
- In response to this new national priority, patient safety activities were undertaken by numerous organizations, including federal agencies, state governments, state coalitions, health care providers, professional associations, and other private organizations.

Implications for the AHRQ Patient Safety Initiative at Baseline

In *Evaluation Report I*, we identified the following implications for AHRQ as it implements the patient safety initiative, which continue to be relevant in 2005:

- *AHRQ leadership*—a clear mandate by Congress for AHRQ to provide leadership in effecting change in patient safety practices.
- *Balance between research and implementation*—the need for AHRQ to balance its traditional role of funding health services research with a shift to an implementation focus to catalyze patient safety improvements in the health care system.

---

[1] The QuIC was composed of members representing the Departments of Commerce, Defense, Health and Human Services, Labor, State, and Veterans Affairs; the Federal Bureau of Prisons; the Federal Trade Commission; the National Highway Transportation and Safety Administration; the Office of Management and Budget; the Office of Personnel Management; and the U.S. Coast Guard.

- *Resource constraints*—small appropriation of funding relative to the work at hand, including research to strengthen knowledge and actions to disseminate that knowledge to the health care community and increase adoption of safer practices.
- *Accountability for results*—high expectations by Congress that AHRQ demonstrates progress in improving patient safety practice and reducing harm to patients.
- *Coordination of multiple activities*—a diversity of patient safety activities undertaken by multiple public and private organizations, which requires a coordination role for AHRQ to achieve synergy among those organizations and to encourage consistent standards of practice.

Recent Events Affecting the Patient Safety Initiative

In *Evaluation Report II*, we identified two major events with actual or potential effects on the scope of the patient safety initiative: the shift in focus of patient safety appropriations toward health information technology (IT) starting in FY 2004; and the pending passage of legislation that would establish patient safety organizations (PSOs) and create protections for reporting of adverse events to the PSOs. Over the past year, several new developments have raised the level of attention to patient safety, with a particular focus on using health IT to achieve safer care, as described below.

Appropriations for health IT grants. As noted in *Evaluation Report II*, Congress appropriated $50 million in FY 2004 to support health IT projects that improve patient safety and quality in health care, along with $10 million to support health IT standards development, with similar amounts appropriated in two subsequent fiscal years. This funding replaced the previous $50 million in annual appropriations that had supported the first group of patient safety projects. AHRQ funded a total of 108 projects to plan, implement, or assess the value of various health IT applications. Planning grants were for a one-year period; implementation grants and grants that assess the value of health IT applications (called value grants) were for a three-year period. AHRQ also funded a second round of 16 implementation grants in FY 2005, all of which were awarded to organizations that had received health IT planning grants in FY 2004. The total FY 2005 funding for these new implementation grants was an estimated $7 million, and their duration varies from one to three years.

Legislation for PSOs. The Patient Safety and Quality Improvement Act of 2005 (Public Law 109-41) (PSQIA) was enacted in July 2005. The purpose of the law is to (1) encourage the voluntary reporting of medical errors and adverse events by health care providers; (2) enable the development of a national network of patient safety databases; and (3) reduce the incidence of events that negatively affect patient safety. The law's protections for reporting medical errors and adverse events, along with its support for systematic collection and sharing of data on these events, should add further momentum to patient safety initiatives already under way throughout the health care sector. As of September 2005, it is unclear how the national data network will evolve or how many PSOs will decide to report to a national network.

AHRQ has responsibility for implementing the provisions of the new law, including reviewing applicant PSO certification statements and listing those approved by the Secretary, fostering design and operation of a network of patient safety databases, and providing common definitions and formats that may be used by PSOs and providers. According to AHRQ leadership, the new law has generated considerable activity, and numerous organizations have expressed interest in PSO certification. AHRQ staff are now working with DHHS leaders on setting the direction for the program and preparing procedures and criteria related to PSO certification. They are also proceeding with plans for fostering development of a network of patient safety databases. As of September 2005, Congress had not yet appropriated funding to implement the law's provisions, so it is not clear how much funding AHRQ will have to carry out its new responsibilities. Even with adequate funding, AHRQ will be managing a significantly increased workload and new challenges resulting from these responsibilities, while maintaining its already-existing programs.

Other public- and private-sector initiatives to improve patient safety. During the past year, several patient safety initiatives outside of the AHRQ portfolio have gained momentum, including the Surgical Care Improvement Project (SCIP), led by the Centers for Medicare and Medicaid Services (CMS) and a collaborative partnership of public- and private-sector health care organizations; the 100,000 Lives (now the 5 Million Lives) Campaign, led by the Institute for Healthcare Improvement (IHI); and transformation of health care providers led by CMS and quality improvement organizations (QIOs). AHRQ is a partner or collaborator in all of these initiatives, as well as the leader of a new initiative on high-reliability organizations in partnership with health care systems. In addition, the original group of individuals trained in AHRQ's Patient Safety Improvement Corps (PSIC) has initiated activities to improve patient safety within their organizations (see Chapter 5 for details on this program).

Critiques of patient safety improvement progress. Assessments of the status and progress of patient safety in the United States began to emerge in 2004, in anticipation of the fifth anniversary of the IOM's seminal report *To Err Is Human*. Altman, Clancy, and Blendon (2004) expressed concern that improvements in patient safety are being blocked by lack of consensus among policymakers, health professionals, and the public on what systemwide steps need to be taken and which events should be reported. At the November 2004 Quality Improvement Colloquium, sponsored by the Commonwealth Fund, keynote speaker Robert Wachter gave patient safety efforts an overall grade of C+, noting striking areas of progress tempered by clear opportunities for improvement (Wachter, 2004). Leape and Berwick (2005) identified additional barriers to safety improvements, including the complexity of health care, commitment to individual professional autonomy, providers' fears of liability, lack of leadership commitment, paucity of measures, and perverse incentives created by payment policies.

STRATEGIC CONTEXT AND ORGANIZATION

AHRQ Patient Safety Strategy and Goals

This evaluation examines the evolution of the AHRQ patient safety initiative in the context of the agency's overall strategy and goals, assessing the extent to which AHRQ is meeting its own goals as well as identifying other areas in which activities are being undertaken or should be pursued but that are not addressed in its goals. During FY 2004, AHRQ defined a new mission, which moves the agency away from its previous focus on research and toward an explicit commitment to quality and safety in health care through a combination of scientific research and promotion of improvement (AHRQ, 2004b). The strategic plan that guides its activities has four goals: safety and quality, efficiency, effectiveness, and organizational excellence. As reported in the AHRQ FY 2005 budget justification, the patient safety initiative is contributing to the first three overall agency goals (AHRQ, 2004b).

In 2003, AHRQ established a new patient safety strategy, which replaced its initial patient safety ten-year plan. AHRQ is using a four-element framework to structure its long-range work and performance assessment: (1) identifying threats to patient safety; (2) identifying and evaluating effective patient safety practices; (3) teaching, disseminating, and implementing effective patient safety practices; and (4) maintaining vigilance (AHRQ, 2003b). The performance goals and fiscal-year targets for implementing the first three elements of this plan were provided in *Evaluation Report II*. Our evaluation is finding that AHRQ is on track thus far for achieving its stated goals, and our full assessment of its status as of the end of the evaluation will be presented in our Final Report (*Evaluation Report IV*). Our interviews with AHRQ leadership indicate that the agency may reexamine these goals and performance targets in the near future, for which it will have the cumulative findings of this evaluation to help guide any changes it makes.

AHRQ Organization for the Patient Safety Initiative

AHRQ's overall programming is managed by five centers, all of which are involved in the patient safety initiative to varying degrees. The Center for Quality Improvement and Patient Safety (CQuIPS), the Center for Primary Care, Prevention, and Clinical Partnerships (CP3), and the Center for Delivery, Organization, and Markets (CDOM) have been the most actively involved in patient safety activities to date, and they will continue to have lead roles in FY 2006 and beyond. CQuIPS has the primary responsibility for overall management of the patient safety initiative. CP3 has the lead responsibility for awarding and managing the health IT grants.

At the start of the patient safety initiative, AHRQ established the Coordinating Center to serve as a stimulus and facilitator of interactions among the projects funded in FY 2000 and FY 2001. The first contract was awarded to Westat, effective October 2001, with a three-year term that ended in September 2004. The second three-year, $3.75-million contract (with a provision for two one-year options at $1.5 million per year) was awarded to the National Opinion Research Center (NORC). This second contract places much greater emphasis on dissemination and implementation activities, as well as on development of tools and products for the health care community. The new Coordinating Center also supports broader patient safety activities within the agency, including management support for the patient safety portfolio.

In September 2004, AHRQ also awarded to NORC a five-year, $18.5-million contract to establish a National Resource Center for Health Information Technology (Resource Center). The Resource Center provides technical assistance and support for the health IT grantees and assists AHRQ with managing the health IT program.

UPDATE ON AHRQ PATIENT SAFETY ACTIVITIES

Congress has continued to appropriate funds to support AHRQ patient safety grants and activities. The history of funding for patient safety grants is summarized in Table 2.1. AHRQ also has continued to fund other investigator-initiated projects on patient safety issues, which contribute to expansion of the patient-safety knowledge base. We provide brief summaries of these grant programs below; more-detailed information about the projects is provided in Chapters 2 and 4 of *Evaluation Report II* (Farley et al., 2007).

Summaries of Patient Safety Projects

FY 2000 and FY 2001 Patient Safety Grants. A total of 81 projects was awarded AHRQ funding as part of the FY 2000–FY 2001 patient safety portfolio, and CQuIPS was responsible for their overall management. These include six projects under the systems-related best practices (SRBP) Request for Application (RFA) issued in 2000, as well as 75 projects funded under the six RFAs issued in 2001. AHRQ obligated a total of $142 million over the life of the multiyear grants; the reporting demonstrations represent half of the spending.

Table 2.1
History of AHRQ Funding for Groups of Patient Safety Grants, FY 2000–FY 2005

Fiscal Year	Type of Projects	Annual Funding Amount
FY 2000	Systems-related best practices	$2 million
FY 2001	Six groups of patient safety grants	$50 million
FY 2001	Working conditions grants	$7 million
FY 2003	Challenge grants	$4 million
FY 2004	Health IT grants and contracts	$50 million
FY 2005	Partnerships in patient safety grants	$3 million
FY 2005	Health IT implementation grants	$7 million

Patient Safety Projects Funded by HRSA. In September 2001, AHRQ and the Health Resources and Services Administration (HRSA) collaborated to include five additional HRSA-funded projects (for a total of $2.4 million) in the designated patient-safety portfolio. These projects focused on developing and testing methods for interdisciplinary training on patient safety for medical and nursing students.

Challenge Grants for Patient Safety Practices. In FY 2003, AHRQ awarded nearly $4 million for 13 challenge grants, with the grantees providing matching funds. These included seven grants for implementation of proven patient safety practices and six grants to test use of risk-assessment techniques for identifying and reducing patient safety issues in health care organizations. The one-year risk-assessment grants were completed at the end of FY 2004; the three-year implementation grants completed two years of work at the end of FY 2005.

Health IT Grants for Patient Safety and Quality. In FY 2004, AHRQ awarded $139 million in multiyear grants and contracts to implement and evaluate the use of health IT for improving patient safety and quality of care; the funded projects provided matching funds. These projects, collectively entitled "Transforming Healthcare Quality Through Information Technology," were funded through three separate RFAs: planning grants, implementation grants, and grants to demonstrate the value of health IT. In FY 2005, AHRQ awarded an additional $7 million to a new set of 16 health IT implementation grantees who had received planning grants the year before.

State and Regional Demonstrations in Health IT. In FY 2004, AHRQ awarded a total of $25 million in contracts over five years to five states (i.e., Colorado, Indiana, Rhode Island, Tennessee, and Utah) to undertake statewide or regional demonstration projects that utilize health IT to improve data sharing and interoperability among health care providers, payers, purchasers, and other stakeholders. A sixth contract was awarded to Delaware in FY 2005.

Grants for Partnerships in Implementing Patient Safety (PIPS). In June 2005, AHRQ awarded $9 million under the cooperative agreement mechanism to fund 17 PIPS grants for up to 24 months. The overall goal is for institutions to work in collaboration with AHRQ to implement safe-practice interventions designed to eliminate or reduce medical errors, risks, and harms associated with the care process. The grantees are required to develop tools as products of their projects and package the tools in an implementation toolkit for use in future implementation projects at their institution. The grantees are required also to work with AHRQ on dissemination of their project results and products. Although AHRQ encouraged applicant institutions to make a substantial commitment of support by providing resources to the projects, it did not require a specified matching-funds ratio.

AHRQ Leadership for Federal Patient Safety Activities

Although AHRQ is the only federal agency that has received substantial funding specifically for patient safety work, other agencies have become increasingly involved in implementation of patient safety improvements in the field. These include CMS, the Centers for Disease Control and Prevention (CDC), the Food and Drug Administration (FDA), and HRSA.

The Patient Safety Task Force (PSTF), which was established in 2001, was charged to develop an integrated data system for patient safety data reported to the DHHS agencies. AHRQ funded and directed the integrated-database project on behalf of the PSTF, until it was discontinued at the end of FY 2004. Integration of various agencies' patient safety systems and databases proved difficult, in part because of the very different mandates, objectives, operations, and constituencies addressed by each agency. (This experience is assessed as part of the monitoring system discussion in Chapter 3.) The PSTF has not undertaken any other significant patient safety activities.

As part of its broader role within the national patient safety infrastructure, AHRQ also provided leadership and operational and financial support for the QuIC. In 2004, the QuIC ceased activities, in part because many of the joint-agency projects it sponsored were not relevant for many of its members. Our observations indicate that the discontinuation of QuIC did not have a substantive effect on ARHQ's patient safety activities.

Financial Resources and Budgets

As shown in Figure 2.1, patient safety funding has been stable or increasing since the start of the initiative. Funding increased from 18.4 percent of the AHRQ budget in FY 2001 to 26.3 percent in FY 2005. If Congress approves AHRQ's proposed FY 2006 budget, patient safety will remain at 26.3 percent of the total AHRQ budget, with $60 million of the $84 million in appropriations allocated for health IT projects.

PREPARATION FOR IMPLEMENTING PATIENT SAFETY ORGANIZATIONS

With enactment of the PSQIA in July 2005, AHRQ was designated as the implementing agency and CQuIPS had responsibility for carrying out the program. As of September 2005, AHRQ staff were working with DHHS leadership on setting the direction for the program and preparing the procedures and criteria needed for certification of PSOs. This program is a regulatory function that differs from AHRQ's usual functions; AHRQ and DHHS are drawing upon expertise in regulatory agencies to help inform their program design and operation.

AHRQ leaders expect the PSO program to greatly increase its workload, with a consequent need for staff and funding to implement the program effectively. Even though the PSO certification process is largely attestation by the applicant PSO that it meets published requirements, AHRQ is responsible for ensuring that entities listed by the Secretary of DHHS are performing effectively. As of September 2005, it was unclear how the national data network would evolve. AHRQ staff are well aware of the complexity of the technical and procedural issues involved.

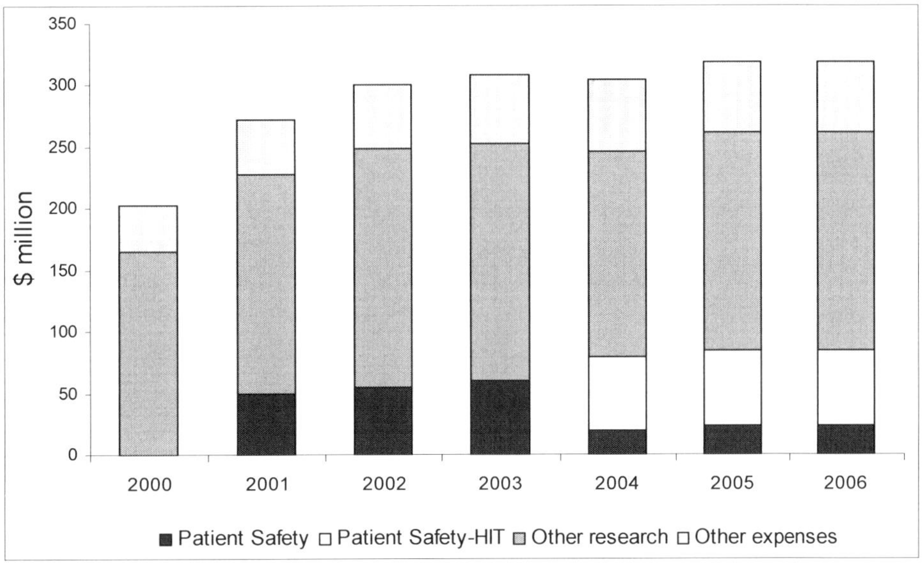

Source: DHHS Budget in Brief, fiscal years 2002, 2003, 2004, 2005, and 2006.

Note: Other research areas include Translating Research Into Practice (TRIP), Consumer Assessment of Health Plans (CAHPS), Healthcare Cost and Utilization Project (HCUP), and other quality, cost-effectiveness and intramural research. Other expenses include Medical Expenditures Panel Survey (MEPS), Current Population Survey, and program support.

Figure 2.1 Trends in AHRQ Budgets for Patient Safety and Other Functions, FY 2000–FY 2006

ISSUES TO CONSIDER

From the beginning, the AHRQ patient safety initiative has been a model for building multiyear strategies in which investments in knowledge-development research are explicitly linked to subsequent practice-improvement activities. As the initiative moved forward in FY 2004–FY 2005, AHRQ expanded its focus to include an emphasis on dissemination activities and support of implementation-oriented projects. At the same time, large initiatives led by other organizations were reinforcing the diffusion of proven patient safety practices in the field.

Within this policy and operational context are several categories of actions that AHRQ should consider as it directs the future of the initiative. These categories emerged from assessments performed in the evaluation. Many of these issues and actions have been identified in previous evaluation reports, but our consideration of them has been modified to reflect changes in AHRQ's emphasis and strategies as its patient safety initiative has evolved. We summarize each action here, referencing later chapters with relevant assessment results.

Proactively Disseminate Patient Safety Practices

This is the time for organizations committed to patient safety to move aggressively to encourage and support health care providers in adopting improved practices for safer health care. Leaders in the field are sensing that policymakers may soon become impatient with the shortage of measurable improvements, and enthusiasm for the patient safety activities may begin to erode.

Yet, the research and development work of the past few years, much of which AHRQ has funded, has begun to yield information and products that are being used by some exciting new national-level safety-improvement initiatives led by various organizations. We encourage AHRQ to continue to accelerate its work to identify priorities for additional patient safety practices, and to generate and disseminate the products and tools to support end users in adding these new practices to their operations. (This issue is the subject of Chapter 6 of this report.)

Collaborate in Initiatives Sponsored by Others

Since *Evaluation Report I*, we have emphasized the need for AHRQ to be strategic in selecting the activities it is best suited to perform as a governmental agency and in identifying activities best performed by others or through partnerships. The recent emergence of new patient-safety-partnership activities highlights the power of collaborative efforts led by field-based organizations. AHRQ partners in many of these initiatives, contributing to them from its relative strengths, such as identifying new safety issues or practices through its funded research and packaging selected products for provider use. AHRQ also is reaching out to providers to test new ways to collaborate with end users on implementing safer health care systems, which should yield valuable insights that the agency can apply in other dissemination activities. However, demand from the field could exceed AHRQ's resource capacity, requiring careful choices to maximize its leverage for improving safety outcomes (see Chapter 6).

Focus on a National Patient Safety Data Capability

The intersection of several recent events provides AHRQ with a unique opportunity to make inroads in establishing a national patient safety data capability—a priority that we have emphasized since *Evaluation Report I*. These events include the passage of the PSQIA, which provides for voluntary reporting to a national data network, the large portfolio of health IT projects funded by AHRQ, the growing availability of components needed for such a system, establishment of regional health information organizations, and increased federal emphasis on electronic health records (EHRs) and achieving information system interoperability. AHRQ is contributing to expansion of EHRs through its funding of the health IT projects, many of which are implementing EHRs or similar record systems (see Chapter 4). To build a national data capability, data for key measures on safety practices and outcomes will be needed, including data from surveys, event-reporting systems, administrative records, and EHRs. In implementing the PSO certification and data networks under PSQIA, AHRQ has the opportunity to accelerate collaborative work on data-system reporting processes and standards (see Chapter 3).

Plan Future Patient-Safety-Funding Priorities

Evaluation Report II addressed the tension involved in allocating AHRQ's finite funding between its patient safety research and implementation functions. As efforts are made to move the results of completed projects into the hands of end users, the emphasis understandably has shifted toward dissemination and adoption. By the time the results from the FY 2000 and FY 2001 projects are packaged and disseminated and the health IT projects have neared completion, AHRQ should have established priorities for moving forward with future phases of its work.

In identifying future directions for changing patient safety practices and culture, AHRQ can draw upon information from this evaluation regarding which safety areas require more research or field testing (Farley et al., 2005; Farley et al., 2007). It also can gain valuable input from end users on their needs for additional resources or knowledge. AHRQ should consider holding a series of regional meetings, with participation by various stakeholders, to help identify and rank priorities for patient safety research or actions. By responding to end users' priorities, the products of AHRQ's future work will be more useful to the audiences it is intended to help.

CHAPTER 3. PROCESS: MONITORING PROGRESS AND MAINTAINING VIGILANCE

Establishment and monitoring of indicators to assess performance improvement progress for key patient safety processes or outcomes, while maintaining continued vigilance to ensure timely detection and response to issues that represent patient safety risks and hazards.

This chapter focuses on the first of five system components of an effective patient safety system, as depicted in Figure 1.1 of this report. This component is monitoring progress and maintaining vigilance, as defined in the box above.

BUILDING FROM *EVALUATION REPORTS I* AND *II*

The ability to assess performance on established patient safety measures is vital to the health care community for reducing risk of harm to patients and the consequences of such harm. This performance information will also enable AHRQ to report to Congress on the results of its investment in patient safety research and dissemination so that appropriate adjustments can be made to the patient safety portfolio.

In *Evaluation Report I*, we examined the key components needed for success in developing a national-level repository for patient safety data. We also assessed the current status of data availability and use of patient safety standards in the field. We concluded that establishing a national-level data repository may be one of the most important steps for the national patient safety initiative, as well as one of the most difficult. In *Evaluation Report II*, we examined progress toward establishing a national data repository, and we reiterated our conclusions about the importance and difficulty of achieving this goal.

During FY 2004–FY 2005, AHRQ modified its approach and activities for development of a patient safety data repository and measures, drawing on experiences from earlier projects as well as on health IT activities in the field. Several AHRQ-funded projects that were developing key components for such a system completed their work, and the experiences and products from those projects informed AHRQ's new participative approach to developing a network of databases. At the same time, activity by both public and private organizations increased on health information technology initiatives, creating a growing consensus on the need for standardization of data reporting and other aspects of systems to support clinical care processes. In this chapter, we document these recent activities and offer suggestions to AHRQ for building upon these developments. Our evaluation addresses the following research questions, most of which were also identified in *Evaluation Report II*:

- What progress has been made in establishing and using national standards and measures for achieving consistency in patient safety reporting systems?
- What actions have been taken to establish a national-level patient safety data repository based on consistent standards for data being reported and system design?
- To what extent are national-level data available regarding the performance of the U.S. health care system on patient safety measures, and how has this changed since last year?

- How has the status changed since last year in the use of generally accepted patient safety measures for assessing performance as part of accreditation or other credentialing processes?
- What steps need to be taken to enhance the capability for effective monitoring of patient safety performance?

Our assessments draw on information provided by a variety of data sources, including relevant written materials and documents, interviews with AHRQ staff, other federal agency staff, representatives of state reporting systems, and principal investigators (PIs) for the AHRQ-funded projects.

STANDARDS FOR PATIENT SAFETY REPORTING SYSTEMS

AHRQ has established a goal to develop a reporting capability through a national-level patient safety data repository or network of databases, to be reached by supporting work on data standards and building capability for aggregation of data from local, regional, and state levels to the national level. To this end, AHRQ has now completed three projects as described below.

IOM Project on Patient Safety Data Standards

The 2004 IOM report *Patient Safety: Achieving a New Standard of Care* presented a plan to facilitate the development of data standards applicable to the collection, coding, and classification of patient safety information, along with recommendations for standards to be adopted by health information and reporting systems. The first recommendation focused on the need for comprehensive patient safety systems in health care organizations to provide immediate access to complete patient information, captured as a by-product of care, and decision support tools for clinicians and patients. The IOM then called for development of a national health information infrastructure (NHII), through which data can be exchanged and communicated across these local data systems to support safety improvements (IOM, 2004). These recommendations are now helping to guide federal government and private-sector activities for developing standardized information systems and interoperability among them.

Federal Data System Project

The federal data system project, sponsored by the DHHS Patient Safety Task Force, sought to create an adverse-event-reporting interface system that integrates reporting for the CDC National Healthcare Safety Network and the FDA adverse-event-reporting systems. This integration effort was to be the first phase in a strategy that would develop a common model for additional agencies and, ultimately, for other organizations across the country. The federal data system consisted of two components: a Web-based event-reporting screen that interfaces with the user, and a data warehouse in which the data resides. Most of the development work was completed by September 2004. However, neither the FDA nor the CDC planned to move forward with the integrated system, owing to very different mandates, objectives, operations, and constituencies addressed by each agency.

AHRQ staff noted that the initial focus of the federal data system project was on the technology, with less attention paid to the needs of the users, in particular, their sense of ownership of the product. These concerns are germane to many information-technology-development efforts. AHRQ has modified its approach to this work so it now is (1) focusing first on the users' information needs, with a recognition that such a broad diversity of needs can be met only through a comprehensive, flexible design, and (2) planning to establish an open network of interoperable databases from multiple sources.

AHRQ's efforts to test the system's application with other organizations through an advisory panel of end users revealed that the system was too heavily oriented toward federal-agency data systems to be of use to others. AHRQ decided to move forward with collaborative development of standards and design for use in any patient safety database, working with end users and other stakeholders to focus explicitly on their diversity of information needs.

The Joint Commission Taxonomy for Patient Safety Reporting

Funded in part by AHRQ, the Joint Commission project sought to establish a common taxonomy for patient safety reporting, working with stakeholders who provided input on users' needs. The Patient Safety Event Taxonomy (PSET) was designed to enable interoperability of reporting systems and information comparability across systems and over time. Drawing upon existing taxonomy models, the PSET identified five unique primary classifications (i.e., impact, type, domain, cause, and prevention and mitigation) that were then divided into subclassifications, coded categories, and noncoded text fields (Joint Commission, 2005; Chang et al., 2005). In August 2005, the National Quality Forum (NQF) endorsed this taxonomy, along with standard reporting elements for patient-safety-reporting systems, guiding principles to improve the taxonomy, and recommendations on the role of the taxonomy in the health IT infrastructure (Kizer et al., 2006).

CURRENT ACTIVITIES FOR STANDARDIZED DATA SYSTEM

Establishment of consistent standards for health information systems and measures encompasses two general categories of systems: institutional systems adopted by local health care organizations (e.g., EHRs); and reporting systems at regional, state, or national levels through which data from multiple health care organizations can be aggregated. DHHS and private-sector activities are currently under way to address consistency in standards and provide support for system infrastructure for both categories of information systems. AHRQ is supporting the development of local and regional information systems through its health IT project portfolio, as well as the establishment of consistent standards for aggregation of reporting data from providers to external data networks.

AHRQ Development of a National Data Network

A plan for development of a national network of patient safety databases was developed by AHRQ in early 2005, reinforced by the passage of the Patient Safety and Quality Improvement Act of 2005. This plan draws upon insights gained from the various projects funded by AHRQ, as described above. The goal of this effort is to develop a taxonomy and data definitions for consistent reporting of patient safety events across the country, along with a data warehouse in which reported data can be aggregated at the national level. Judging from recent research results from an AHRQ-funded project, we note that common definitions may not be sufficient to enable meaningful aggregation of data because individual providers (or units within larger organizations) vary widely in how they interpret and classify similar events (Tamuz and Thomas, 2006). As of September 2005, this work was in the very early stages of discussions with stakeholders, and decisions had not yet been made regarding how the data would be used, who the users would be, and under what conditions they would have access to the data.

To facilitate rapid progress in database design, AHRQ is building upon products and information that are already in place, including the Joint Commission PSET reporting taxonomy, work by RAND on documenting the scope and data elements of state reporting systems, and the hospital survey of adverse-event-reporting systems that is being performed as part of this evaluation, in collaboration with Joint Commission. In addition, the Coordinating Center contract has been modified to create a temporary repository at NORC for an inventory of reporting systems.

DHHS Work on Interoperable Health Data Systems

The DHHS Office of the National Coordinator of Health IT (ONC), established by Executive Order 13335 in April 2004, is charged with coordinating all federal initiatives related to health IT and developing "a strategic plan to guide the nationwide implementation of interoperable EHRs in both the public and private health care sectors" (ONC Web site, 2005). The plan, released in July 2004, lays out four broad goals for a national health IT agenda: (1) informing clinical practices by providing them more information at the point of care; (2) interconnecting clinicians by establishing an interoperable health information infrastructure; (3) personalizing care by giving consumers a greater role in making decisions concerning care; and (4) improving population health. The plan also presents a strategic framework with specific strategies for achieving these goals, ranging from fostering regional health information organizations to developing a national health information infrastructure to promoting the adoption of personal health records (Brailer, 2004).

Also in 2004, DHHS convened a Health IT Leadership Panel to investigate the costs and benefits of health IT and devise recommendations for maximizing the benefit-to-cost ratio for industry and society. The panel determined that the potential benefits of health IT are much higher than the costs, and recommended that "widespread adoption of interoperable health IT should be a top priority for the U.S. health care system" (The Lewin Group, 2005).

In June 2005, the Secretary of DHHS formed the American Health Information Community (AHIC), a collaborative of key public- and private-sector experts in health IT to advise DHHS on strategies for pursuing a common interoperability framework, creating a private-sector process for the development of health IT standards, and certifying health IT products. DHHS has also solicited and received input from a wide range of stakeholders in the health care and IT sectors on how best to proceed with building a national, interoperable health data exchange network.

Key industry groups have responded to calls by DHHS and ONC for the private sector to take the lead in developing a process and criteria for certifying health IT products. Several private health IT organizations have formed collaboratives for development of EHRs, testing of methodologies and standards, and laying of the groundwork for development of an NHII. Taken together, these efforts offer an important opportunity for the federal government, including AHRQ, to collaborate with the private sector in achieving the goal of establishing an NHII and, within that, a national patient safety data network. It remains to be seen whether these separate initiatives will build upon each other in a way that substantially reduces the barriers to widespread adoption of health IT and leads to improvements in patient safety and quality of care.

AVAILABILITY OF PATIENT SAFETY MEASURES

The absence of a comprehensive set of national patient-safety process and outcome measures that are generally accepted in the health care community continues to hinder the tracking of patient safety performance. As described in *Evaluation Reports I* and *II*, various sets of outcome measures have been developed by several agencies and private organizations, including AHRQ. However, the established measures address a limited number of health care settings and only selected aspects of care within them. Virtually all measures apply to the hospital inpatient care setting; there are no safety measures for ambulatory care and only a few measures for long-term care. Furthermore, little work has been done to establish process measures for assessing the adoption and actual use of effective patient safety practices.

In *Evaluation Report I*, we suggested that AHRQ use a structured consensus process for identifying a generally accepted set of measures that would represent the most important patient safety issues, with participation of key stakeholder groups in that process. The publication of measures by AHRQ, as the lead agency in the patient safety initiative, would tend to stimulate movement toward consistent use of those measures, as demonstrated by the AHRQ patient safety indicators (PSIs), which have been used in analyses published in numerous papers as well as by a number of hospitals for assessing their own patient safety performance. The PSIs are inpatient safety outcome measures developed by the Evidence-based Practice Center at the University of California, San Francisco, and Stanford University, with collaboration from the University of California, Davis, using hospital inpatient discharge data from the HCUP National Inpatient Sample (NIS) (McDonald et al., 2002).

The NQF consensus process for comprehensive review and endorsement of measures offers the possibility of moving toward establishment of such measures. The NQF established the basis for process measures in 2003, when it endorsed a set of 30 evidence-based safe practices for use in applicable clinical care settings; however, data-collection methods for measuring use of these practices do not yet exist. The only outcome measures the NQF has addressed to date are the 27 serious reportable events that are intended to form the basis for a mandatory state-based reporting system.

The RAND evaluation team is assessing patient safety outcome measures as part of its product evaluation (see Chapter 7). Thus far, we have performed baseline trend analyses on several of the PSIs developed by AHRQ, as well as on other measures developed by two reporting demonstration projects. The PSIs capture events that occur in hospital inpatient services, focusing on complications and adverse events following surgeries, procedures, and childbirth. The results of this analysis should support efforts to establish a national set of patient safety measures for monitoring health care performance trends. Ideally, this evaluation should be able to work with an already-established, comprehensive set of outcome measures to estimate effects of the patient safety initiative. Instead, we are using a limited set of available measures while also “pushing the envelope” by testing analyses of other newly developed measures.

USE OF MEASURES IN ACCREDITATION OR CREDENTIALING

Because accreditation and credentialing organizations set standards and measures for performance of health care providers and related information systems, they can build synergy toward adoption of national patient safety measures and data standards by providing consistent policy direction to providers. Here, we update the information provided about these activities in

Evaluation Reports I and *II*, with a particular focus on Joint Commission and Medicare, the two entities with the most-active programs.

The Joint Commission

The Joint Commission has addressed patient safety in its accreditation process since 1996, with the establishment of its sentinel-event policy,[2] followed by a comprehensive patient safety policy. The Joint Commission now includes standards for patient safety practices in its accreditation of all health care organizations. To help accredited organizations address patient safety areas of specific concern, the Joint Commission has also established National Patient Safety Goals, which highlight problematic areas and focus on systemwide solutions (Joint Commission, 2005). The patient safety goals have been in effect since the accreditation process in January 2003, and they are updated each year. The Joint Commission is also participating in and supporting numerous efforts to establish additional patient safety measures and measurement capability.

Medicare Program Requirements and Activities

To participate in Medicare, hospitals, other institutional providers, and health plans must meet certain conditions that are intended to protect patient health and safety and to ensure that high-quality care is provided to all patients. The new condition established in 2003 requires hospitals participating in Medicare to develop and implement a quality assessment and performance improvement (QAPI) program that identifies patient safety issues and reduces medical errors. The QAPI requirement continues to be the foundation of the Medicare patient safety standards.

The Center for Medicare and Medicaid Services does not plan to establish standardized patient safety measures for Medicare until a national core set of hospital performance measures is established, including patient safety measures. CMS continues to be an active participant in collaborative activities to develop such measures. Through its Medicare Patient Safety Monitoring System (MPSMS), CMS is generating national estimates of incidence rates for a growing number of patient safety measures, some of which have been included in the AHRQ *National Health Quality Report* (NHQR), which can be used by health care organizations for benchmarking their performance. The first set of MPSMS results was released in 2005, and new measures will be added each year. The MPSMS is a rich source of candidate patient safety measures for consideration in a consensus process.

In recent years, CMS has placed an emphasis on promoting health IT in several of its programs. For example, the CMS Doctors' Office Quality–Information Technology (DOQ-IT) program is encouraging the adoption of EHRs in small- and medium-sized primary care physicians' offices. A two-year demonstration program in four states (i.e., California, Arkansas, Massachusetts, and Utah), DOQ-IT provides technical assistance and support, largely through QIOs, to physicians' offices that are in the process of adopting EHRs and related health IT, to improve treatment of Medicare beneficiaries with chronic conditions. CMS is also working with QIOs to promote the adoption of other kinds of health IT in a variety of health care settings.

---

[2] *Sentinel events* are the most serious adverse events.

ISSUES AND ACTION OPPORTUNITIES

AHRQ faces both opportunities and challenges in bringing about a national patient safety data network. The passage of the Patient Safety and Quality Improvement Act of 2005 has shifted the balance toward opportunities, because it provides for development of a network of databases as part of the larger PSO program, which offers confidentiality protections that should encourage reporting by providers. Challenges remain, however, especially in the need to reach consensus among stakeholders on both the measures being entered into the data system and the standards for that data system.

Given the active involvement of the federal government and growing numbers of private-sector organizations in health IT, we can expect that the development process will be fluid, with numerous false starts and restarts. Although the 2004 IOM report offers a useful starting point for standards development, the responsibility for providing direction and leadership so that one set of standards can emerge will probably rest with federal agencies. AHRQ's leadership can be applied to stimulate development, dissemination, and adoption of these standards, including work with end users to ensure that the system designs are serving their needs.

Issues to Consider

AHRQ's decision to modify its data-repository-development process to be more participative and use more-open structures is a reasonable response to the experiences and lessons learned from its earlier projects. Rather than its initial focus on the technology of integrating data systems of federal agencies, it now is focusing first on the information needs of multiple potential users, recognizing that such a broad diversity of needs can be met only through a comprehensive, flexible design.

We encourage AHRQ to remain flexible in seeking ways to establish a viable national patient safety data repository, responding to new methods or tools that arise in the field of health IT and making full use of relevant products from the earlier AHRQ-funded projects—in particular, the reporting demonstrations. AHRQ also should place a priority on further development of patient safety measures that could apply to aspects and settings of care for which measures do not yet exist—in particular, ambulatory care and long-term care settings. With these considerations in mind, we offer the following suggestions for AHRQ action with respect to monitoring progress and maintaining vigilance.

Suggestions for AHRQ Action

- **AHRQ should continue to pursue the goal of developing a national-level patient safety data capability in which multiple public and private users participate, with reinforcement by provisions in the newly enacted patient safety legislation for voluntary reporting by patient safety organizations into a network of patient safety databases.**

The establishment of a national patient safety data repository is essential for enabling effective tracking across the country of valid and reliable measures of patient safety processes and outcomes and for eliminating the proliferation of independent reporting systems that providers find burdensome. The formation of such a system depends on having national consistency in data system structures, variable definitions, and data standards. As described in this chapter, AHRQ has undertaken several activities and contracts in the past few years to define measures and build the core elements of a national patient safety data repository. The technical products from this work should be used whenever appropriate and feasible.

A major barrier to potential progress has been concern by providers that they would have medical liability exposure if they reported data on errors and adverse events to external organizations. This barrier was eased by the passage of PSQIA in July 2005, which provides protections for reporting data to PSOs and permits submittal of the data to a national data repository. With this legislation in place, continued leadership by AHRQ in facilitating a consensus approach to design decisions will be essential for developing a viable national data repository. We encourage AHRQ to work collaboratively, consulting with participating organizations to learn what will motivate them to report data into a national repository, identify concerns that might cause them to not participate, and develop standards for data and reporting that are acceptable to users.

- **Using a structured consensus process involving multiple stakeholders, AHRQ should place a priority on building on the existing Patient Safety Indicators to establish a broader set of national patient safety measures that represents the most important safety aspects of the patient's health care experience in a variety of settings.**

Defining a national set of patient safety measures goes hand in hand with establishing a national data repository, because it is these measures that populate the repository. To fulfill users' information needs, the data in the repository must address clinically important safety issues. Continued measure development by AHRQ and several other organizations, as well as the outcome assessment work in this patient safety evaluation (refer to Chapter 7), are contributing to the measure development process. However, agreement on an expanded set of patient safety measures ultimately is a policy action that is yet to be accomplished.

This suggestion for further work by AHRQ on development of patient safety measures was presented in *Evaluation Reports I* and *II*. We repeat it here because there is a continued need to establish national patient safety measures that extend the scope of health care services and settings being addressed. We again suggest use of a consensus process to facilitate identification and selection of priority measures. To trigger the NQF consensus process, a proposed set of measures must be developed and submitted for NQF review. By including consensus steps in the initial measure-development process, similar to the process used by the University of California, San Francisco (UCSF) and Stanford University to select the PSIs, AHRQ will help ensure that newly-developed measures are valid from the perspective of clinicians and other stakeholders—evidence that will be needed for the NQF review.

CHAPTER 4. PROCESS EVALUATION: KNOWLEDGE OF EPIDEMIOLOGY AND DEVELOPMENT OF EFFECTIVE PRACTICES

> ***Knowledge of Epidemiology of Patient Safety Risks and Hazards:*** *Identification of medical errors and causes of patient injury in health care delivery, with a focus on populations that are vulnerable because they are compromised in their ability to function as an engaged patient during health care delivery.*
>
> ***Development of Effective Practices and Tools:*** *Development and field testing of patient safety practices and tools to identify those that are effective, appropriate, and feasible for health care organizations to implement taking into account the level of evidence needed to assess patient safety practices.*

This chapter focuses on the second and third of the five system components of an effective patient safety system, as depicted in Figure 1.1. These two components contribute to development of knowledge regarding patient safety epidemiology and effective practices and tools, as defined in the box above.

BUILDING FROM *EVALUATION REPORTS I* AND *II*

In *Evaluation Report I*, separate chapters addressed the two components encompassed in the *Knowledge Development* portion of the patient safety system defined in Figure 1.1: (1) the epidemiology of patient safety risks and hazards and (2) establishment of effective patient safety practices and tools. We combined these two components into one chapter in *Evaluation Report II* and continue to do so in this report. This packaging of these two aspects of *Knowledge Development* reflects a shift in emphasis of the overall patient safety initiative, away from a focus on changes in knowledge of epidemiology (what the safety issues are) toward expanded work in developing and testing new safety practices (how to improve safety).

In the epidemiology chapter of *Evaluation Report I*, we assessed the baseline level of knowledge of patient safety epidemiology and relevant issues, through a review of the published literature current at that time. This baseline information was developed to inform our recommendations to AHRQ about further work needed to address epidemiological issues and to begin to track over time the extent to which AHRQ-funded projects were generating published papers that contribute to epidemiological literature. In each evaluation report, we have updated the counts of articles on safety epidemiology published each year, the issues they addressed, and the portion generated by AHRQ-funded projects.

In the safety practices chapter of *Evaluation Report I*, we profiled the initial patient safety projects funded by AHRQ in FY 2000 and FY 2001, including tabulations of the issues they examined, the practices they developed or tested, and the settings for which they were generating knowledge. We also documented the extent to which these projects had the potential to expand the evidence base on effective practices because they were examining practices for which additional evidence was needed, as characterized in the evidence report *Making Health Care Safer: A Critical Analysis of Patient Safety Practices* (Shojania et al., 2001). Finally, we provided information on the experiences of these project teams in carrying out their research, obtained from interviews with project PIs. In each subsequent report, we have focused on the set(s) of projects that AHRQ had funded in the most recent year, updating information on how each new set of projects was contributing additional knowledge on safe practices.

This chapter of *Evaluation Report III* contains updated information on publications addressing the epidemiology of patient safety risks and the contributions of AHRQ-funded projects to this information. It also focuses on three sets of new health IT grants funded in FY 2004, totaling 104 projects. It first presents an update on published information on patient safety epidemiology, and then provides a summary update of the current status of all the sets of patient safety grants funded by AHRQ since the start of the initiative. Then the bulk of the chapter is dedicated to characterizing the health IT projects, assessing how they are contributing to scientific evidence on effective practices and discussing early lessons emerging from their work. The following questions are examined in this third evaluation year:

Epidemiology:

- What additional information has been published about patient safety epidemiology in the past year, and how have the AHRQ-funded research projects contributed to this new information?
- To what extent has the additional research strengthened the evidence regarding epidemiology of errors and adverse events, and in what areas might more research be appropriate?

Understanding Effective Patient Safety Practices:

- What additional research and field tests on new patient safety practices and tools are being conducted by AHRQ-funded projects, and how are they contributing new knowledge regarding practices for which further scientific evidence is needed?
- What range of health IT applications is being tested and studied by the AHRQ-funded health IT projects, and to what extent are those applications addressing patient safety issues?
- What progress has been made by the AHRQ-funded projects in documenting the effects of new patient safety practices and tools on patient safety outcomes and the costs of care, the effectiveness of care for given costs (cost-effectiveness), and return on investment?
- What lessons were learned from the health IT projects regarding effective planning and implementation of health IT, particularly regarding applications to improve patient safety?

EPIDEMIOLOGY OF PATIENT SAFETY

In *Evaluation Reports I* and *II*, we documented the status of knowledge of the epidemiology of patient safety risks and hazards, and identified the extent to which the AHRQ-funded projects were contributing to that knowledge each year. Below, we update this information by including the additional contributions of AHRQ-funded projects to the epidemiology literature during the year ending June 2005. We note that this assessment is limited to identifying the increase in published papers from AHRQ-funded projects as *potential* contributions to the growth in epidemiology knowledge. The *actual* contributions of these projects to epidemiology knowledge can be determined only through a thorough review of the newly published papers, to assess the integrity of their methods and findings, and a synthesis of findings across papers to update the epidemiology evidence base and assess its implications for future patient safety priorities. Such work will require substantial resources and is beyond the scope of this evaluation.

Based on a MEDLINE database search of articles published in English between July 2004 and June 2005, Table 4.1 presents counts of articles addressing epidemiology for a number of patient safety issues. One hundred and fifteen articles, including 10 reviews, were published, providing information on the rates, types, or causes of medical errors or adverse events, which is a substantial increase over previous years. These articles addressed a total of 127 patient safety issues, or 73 percent of what had been addressed in all the previous 4.5 years together. Consistent with previous findings, the largest share of publications focused on medication errors, particularly in relation to the elderly and children, and on general patient safety. Working conditions and surgical-procedure or invasive-procedure errors also received increased attention.

There was substantial growth in the past year in the number of products from the patient safety grantees, which published 31 articles addressing 37 patient safety issues, compared with only four articles in the preceding year. We did not identify any articles generated by the challenge grants funded in FY 2003, which is not surprising, given that their research was still in process. However, two patient safety epidemiology articles were published by grantees with both challenge grants and other patient safety grants, and five articles were published by grantees who also have health IT grants. In addition, four articles were identified as being collaborations between multiple AHRQ-funded patient safety grantees.

Table 4.1
Patient Safety Epidemiology Information Available from Recently Published Articles and Addressed by AHRQ-Funded Patient Safety Projects, Through June 2005

Patient Safety Issue	Articles published 2000 – June 2004	New articles published July 2004 – June 2005	New articles published by AHRQ patient safety grantees	New articles published by challenge grants
Medication ordering, administration	68	55	11	0
Nosocomial infections	3	2	1	0
Falls, pressure ulcers, restraint-related	6	0	0	0
Nurse staffing	6	2	1	0
Provider fatigue, working conditions	4	9	5	0
Surgical or invasive-procedure errors	8	9	1	0
Diagnostic or treatment errors	37	9	1	0
General patient safety	38	34	14	0
Other issues	5	7	3	0
Total number of topics addressed	175	127	37	0
Average number per article		1.1	1.2	NA

NA = not applicable

UPDATES ON GROUPS OF PATIENT SAFETY GRANTS

In this section, we summarize the current status of all the groups of patient safety grants funded by AHRQ since the start of the initiative, as well as our evaluation activities with them. The PIPS grants and health IT projects are the groups most recently awarded funding. The process-evaluation data collection during FY 2005 focused on characterizing and learning from the first year of operation of the health IT grants, the results of which are presented in the remainder of this chapter. The PIPS grants received their funding in September 2005, so their work and experiences will be examined in the FY 2006 evaluation cycle.

FY 2000 and FY 2001 Patient Safety Grants. Many of the FY 2000 and FY 2001 patient safety grants, scheduled to complete work by the end of FY 2004, received no-cost extensions from AHRQ through the end of September 2005. (The Centers of Excellence, which were five-year projects ending FY 2006, also were active in FY 2005). The Coordinating Center and AHRQ have been working with the grantees to document their research findings and develop products for end users in the health care community. This information provides the content for the next steps in AHRQ's dissemination strategy—performing syntheses of related findings across projects, updating the evidence on patient safety practices, and disseminating the practices into the field (see Chapter 6). No evaluation work was performed with these grantees during FY 2005, but project-end interviews have been planned for FY 2006 with a subgroup of grantees that pursued practice interventions, to document their retrospective assessments of the success of their projects and factors contributing to those results.

Challenge Grants. In FY 2004, AHRQ awarded nearly $4 million to support 13 patient safety projects—called challenge grants—that were to assess patient safety risks or to implement evidence-based safe practices. Six of these grants were one-year risk assessment and reduction grants and seven were two-year practice implementation grants. As of September 2005, the risk assessment grants were completed and the implementation grants were coming to a close. To identify early lessons learned from the 13 challenge grants, we conducted in-person site visits in FY 2004 with the seven implementation grant teams and one of the risk assessment grant teams. We also conducted telephone conference calls with the remaining five risk assessment grantees. Follow-up conference calls will be held with the seven implementation grant teams in October and November 2005 to assess overall lessons and experiences.

Partnerships in Implementing Patient Safety Grants. In June 2005, AHRQ awarded $9 million for 17 two-year Partnerships in Implementing Patient Safety grants, to assist health care institutions in implementing safe practices that may eliminate or reduce medical errors, risks, or hazards. As of September 2005, the PIPS grants had been operating for only three months. Interviews with the PIPS grantees have been planned for the first half of 2006 to document their early implementation experiences.

AHRQ Health IT Initiative. In FY 2004, AHRQ awarded $139 million in grants and contracts to advance the use of health information technology. The funding was awarded through three grant initiatives—38 planning grants, 40 implementation grants, and 26 value grants (to assess the value of health IT applications). In addition, five-year contracts were awarded to statewide health IT networks in Colorado, Indiana, Rhode Island, Tennessee, and Utah. In FY 2005, AHRQ awarded an additional $7 million to a new set of 16 health IT implementation grantees who had received planning grants the year before. As of September 2005, interviews had been completed with all of the one-year health IT planning grantees and approximately half of the implementation and value grantees; the remaining interviews were planned to be completed by the end of 2005. Evaluation of the five state and regional networks was planned to begin in early 2006, after their projects were slightly more than one year old; the results will be reported in our final evaluation report (*Evaluation Report IV*).

PROFILE OF AHRQ HEALTH IT PROJECTS

A central question for the evaluation is how the patient safety projects funded by AHRQ are contributing to building scientific knowledge on patient safety practices. Such knowledge is a key component of the system needed to achieve sustainable improvements in the safety of health care (see Figure 1.1). In *Evaluation Reports I* and *II*, we addressed this question for each group of projects previously funded by AHRQ, using two sets of information drawn from their project proposals. First, we characterized the patient safety issues the projects were addressing, the types of safety practices they were developing or testing, the settings in which they were working, and their focus on special populations. Second, we examined the projects' potential contributions to new scientific evidence on safety practices, by identifying the extent to which they were working with practices for which evidence assessments had shown that additional evidence was still needed.

In this report, we present this same information for the health IT projects funded in FY 2004, to add their contributions to those already documented for previous project groups. In addition, we provide further detail on the nature of the health IT projects to characterize the technologies with which they are working, the numbers and types of partners involved, and their rural or urban locations. This information documents the diversity of projects being pursued as part of the patient safety initiative, and it also can be used to identify where future funding decisions should focus to address areas that are not addressed much by the current projects.

When considering the total work on health IT being conducted in the patient safety initiative, it also is important to document the extent to which IT issues or technologies have been addressed by other patient safety projects that are not part of the health IT groups. Therefore, we follow the profiles of the health IT projects by describing the extent to which the research or development work of other patient safety projects involved IT.

Contributions of Health IT Projects to Patient Safety Practices and Infrastructure

Tables 4.2 through 4.6 present the number of health IT grants by their focus on patient safety or quality and special population, geographic setting, numbers and types of partners, and types and purposes of health IT being addressed, respectively. We note the following key points:

- Many health IT projects are focused on reducing errors associated with medication ordering, care procedures and coordination, and administrative events. With this emphasis on system-level interventions, the projects have not addressed many of the more-specific patient safety issues (e.g., nosocomial infections, working conditions, ordering or administration of blood) that have been addressed by the previous patient safety projects (Table 4.2).
- Many of the planning and implementation health IT projects have placed an emphasis on special populations, particularly those who are health vulnerable and low-income. Fewer of the value grantees have done so (Table 4.2).
- A majority of projects is focused in rural areas, and many involve partnerships between urban and rural-based partners. The number of rural projects reflects AHRQ's emphasis on funding health IT projects that targeted rural settings, and actually exceeds the requirement that 50 percent of the projects be rural in nature (Tables 4.3 and 4.4).
- The projects vary in the number of partners involved, with the implementation grants having the largest number of partners. There is a significant representation of hospitals,

physician practices or clinics, and universities, although a wide variety of other types of partners also is involved (Table 4.5).

- The health IT projects place a heavy emphasis on planning for and implementing EHRs, decision support, order entry, and data exchange, with a strong focus on clinical aspects of care-delivery processes. They also tend to work with multiple technologies and purposes, again in contrast to other AHRQ-funded patient safety projects (Table 4.6).

Table 4.2
Patient Safety Issues and Special Populations Addressed by the Health IT Projects

	Number of Issues by Type of AHRQ Health Information Technology Grants		
	Planning (*n*=38)	Implementation (*n*=40)	Value (*n*=26)
Patient Safety Issue			
Medication ordering/administration	19	27	14
Nosocomial infections	0	0	0
Falls/pressure ulcers	0	2	0
Nurse staffing	0	0	0
Provider fatigue, working conditions	0	0	0
Surgical- / invasive-procedure errors	1	0	0
Diagnostic/treatment errors	10	14	13
Equipment / device failure	0	0	0
Ordering / administering blood	0	0	0
Care procedures and coordination *	30	24	6
Wrong patient / procedure / test	0	0	0
General patient safety	1	3	2
Hand-offs	32	24	6
Other issues	0	2	0
Total number of issues studied	93	96	41
Average number per project	2.4	2.4	1.6
Special Populations			
Elderly	8	6	3
Minority populations	5	8	3
Low-income	14	8	3
Health-vulnerable	13	15	6
Other vulnerable	9	4	0

* *Care procedures and coordination* include errors in the admitting process, such as applying the wrong patient identification bracelet; misplaced documentation, such as "lost" medical records; failure to notify patients of a positive test result; failure to register a patient in the emergency department, resulting in delayed care and adverse outcome.

Many of the health IT projects are addressing care procedures and coordination issues, as shown in Table 4.2. These projects tend to involve comprehensive health IT systems that affect a broad range of these issues. Specifically, we found that care procedures and coordination issues were identified most frequently as a focus by projects that were working with data exchange and information (52 projects), electronic health records (44 projects), decision support systems (36 projects), and computerized provider order entry (23 projects). Other types of health

IT that projects were using to address these issues included data collection and summary, results reporting, communication systems, patient decision support systems, and administrative systems.

Table 4.3
Profile of the Health IT Projects Funded by AHRQ

	Planning (*n*=38)		Implementation (*n*=40)		Value (*n*=26)	
	Number	Percent	Number	Percent	Number	Percent
Patient safety or quality focus						
Patient safety only	0	0	2	5	1	4
Both patient safety and quality	34	87	35	88	22	85
Quality only	4	13	3	8	3	12
Geographic setting *						
Rural	21	55%	14	35%	1	4%
Both urban and rural	13	35	19	46	7	27
Urban	4	11	7	18	18	70

* A *rural-only project* had involvement only by organizations outside of a Metropolitan Statistical Area (MSA), as defined by the U.S. Census Bureau, or outside an urbanized area within an MSA and served only a rural population. A *both urban and rural project* involved a mix of urban and rural organizations or served both urban and rural populations. An *urban-only project* had involvement only by organizations located in an urbanized area within an MSA and served only an urban population.

The differences between AHRQ's rural designations and the RAND results are shown in Table 4.4. All of the projects that AHRQ defined as *rural* also were coded either wholly or partly rural by RAND (light shading in the table). RAND also identified 19 other projects with rural components that had not been identified in the AHRQ designations (darker shading in the table), thus increasing the rural projects from 54 percent (56 projects) to 72 percent (75 projects) of the total projects.

Table 4.4
Comparison of Coding of Rural Health IT Projects by AHRQ and RAND

	AHRQ Designation of projects					
RAND Coding	Nonrural	Rural	Both	Small	Missing	Total
Urban only	12	0	0	16	1	29
Both urban and rural	15	20	1	1	1	38
Rural only	2 *	35	0	0	0	37
Total	29	55	1	17	2	104

* Review of information in the proposals for these two projects verified that they are completely rural projects.

To enable comparisons of the newly funded health IT projects with previously funded patient safety projects regarding their uses of IT, we extracted additional data from the proposals for the earlier safety projects about whether their work addressed health IT and, if so, what technologies they considered and for what uses. We show the comparisons of technology and uses for the three groups of health IT projects and other patient safety projects in Table 4.5 (counts of previous projects addressing health IT are given in Table 4.7). With an emphasis on planning for and implementing EHRs, decision support, CPOE, and data exchange, the health IT projects tended to be using a systems approach to address clinical aspects of care delivery—in particular, assessment, documentation, decisionmaking, orders, and cross-cuts (i.e., include clinical and administrative). By contrast, most of the earlier patient safety projects tended to work with more-specific technologies (e.g., mobile computing, data collection and summary).

Table 4.5
Number and Types of Partner Organizations for the AHRQ-Funded Health IT Projects

	Number of Projects by Number and Type of Partner		
	Planning	Implementation	Value
Number of partners			
None	5	2	13
Two	0	1	7
Three	9	7	5
Four	5	10	1
Five to nine	14	18	0
Ten or more	5	2	0
Types of partners			
Networks (loose affiliation)	12	8	2
Individual hospitals	26	25	6
Systems (common management)	5	15	11
Physician practices or clinics	16	22	2
Nursing homes	3	4	1
Foundations	0	4	0
Health Plans	1	5	0
Regional Coalitions	6	4	1
Local or state health departments	7	6	0
University	9	18	10
Quality Improvement Organizations	0	3	0
Other*	10	23	5

* Includes pharmacies, membership organizations, IT companies, independent labs, community-based diagnostic centers, tribal organizations, accrediting organizations, purchasers, behavioral health organizations, home health agencies, hospice, social service agencies, and private research organizations.

Table 4.6
Technologies Being Addressed by the Health IT Projects and Other Patient Safety Projects

	Technology Addressed by IT Grants			Other Patient Safety Projects
	Planning	Implementation	Value	
Type of health IT addressed				
Computerized provider order entry	11	15	9	7
Electronic health records	21	22	2	3
Decision support	18	29	18	19
Results reporting	6	5	3	0
Electronic prescribing	4	4	1	2
Mobile computing	1	6	2	12
Data exchange and information	32	24	3	1
Patient decision support	3	3	5	0
Communication systems	8	7	7	1
Administrative	3	4	0	0
Knowledge-retrieval systems	2	0	1	5
Data collection and summary	9	13	4	16
Other	1	1	0	7
Not specified	1	0	1	0
Total frequency of IT types	120	133	56	73
Average number per project	3.2	3.3	2.2	1.6
Purpose of the health IT				
Clinical operations	4	5	0	2
Clinical assessment	27	27	7	0
Clinical decisions	17	27	19	19
Clinical orders	13	16	9	9
Clinical actions, therapy	0	0	0	3
Clinical documents	28	25	8	3
Clinical cross-cuts *	35	27	14	3
Clinical process improvement	7	12	2	20
Administrative and financial	7	6	0	1
Other or not specified	1	1	1	6
Total frequency of purposes	139	146	60	66
Average number per project	3.7	3.7	2.3	1.4

* *Cross-cuts* are purposes that do not neatly fall into clinical or business purposes, including communication between providers, communication between the provider and patient, managing patient safety, protecting patient confidentiality, managing liability risk, and providing care efficiently.

Consideration of Health IT by Other Patient Safety Projects

As Table 4.7 illustrates, many of the earlier patient safety projects worked with IT solutions for the issues they were addressing. Overall, 49 percent of the projects dealt with health IT, although the percentages varied by project group. By definition, *health IT* was the subject for 100 percent of the clinical informatics to promote patient safety (CLIPS) projects. Among the other groups, the highest percentage was for the implementation challenge grants (86 percent) and the lowest was for the developmental center grants (17 percent).

Table 4.7
Other Patient Safety Projects Addressing Health IT

Project Group	Number of Projects in Group	Projects Addressing Health IT	
		Number	Percentage of Total Group
Systems related best practices	6	2	33%
Reporting demonstrations	16	11	69
CLIPS	11	11	100
Working conditions	21	6	29
Centers of Excellence	3	1	33
Developmental Centers	18	3	17
Dissemination and education	6	3	50
Challenge – implementation	7	6	86
Challenge – risk assessment	6	3	50
All projects	94	46	49%

New Contributions of Health IT Grants to Evidence on Practices

In July 2001, the evidence report *Making Health Care Safer: A Critical Analysis of Patient Safety Practices* was published by UCSF–Stanford University, one of the AHRQ-funded Evidence-based Practice Centers (EPC) (Shojania et al., 2001). This report evaluated a total of 79 patient safety practices, establishing ratings of the strength of evidence for each practice's effectiveness in improving the safety of health care. These practices were categorized into five groups for strength of evidence of effectiveness: greatest, high, medium, low, or lowest. The report also organized the practices into two research groups: "research likely to be highly beneficial" and "research likely to be beneficial" (Shojania et al., 2001). Many potentially effective practices could not be assessed, however, because the necessary scientific evidence from studies of the practices did not exist or was inadequate.

The practices that this initial evidence report identified as needing more evidence to be assessed represent opportunities for development of new evidence on effectiveness. To the extent that AHRQ-funded patient safety projects are addressing these practices, the projects have the potential to contribute strongly to the growth of knowledge on safe practices. Therefore, our evaluation has been assessing how many of the patient safety projects have been working with practices that the evidence report identified as requiring more scientific evidence. Of course, the actual contribution of these projects to science will depend on the integrity of their research methods and findings, which will need to be assessed in future reviews of their published results.

When we examined previous patient safety projects, as reported in *Evaluation Reports I* and *II*, we found that those projects were addressing many practices for which scientific evidence did not yet exist or was weak. In this report, we report results of the same analysis done for the health IT projects. We found that the health IT projects also have the potential to make new contributions to the science of patient safety practices. These results are presented in Table 4.8, which shows the number of health IT projects that address practices at each level of strength of evidence rated by the evidence report. Practices for which there is only medium strength of evidence are being addressed by 37 of the projects (12 planning, 15 implementation, and 10 value), and practices with low or lowest strength of evidence are being addressed by 18 projects. In addition, 53 of the projects are addressing practices for which the evidence report indicated that further research would be beneficial or highly beneficial. Finally, 53 projects are addressing practices that the evidence report did not rate at all.

Table 4.8
Number of Health IT Projects Covering Practices for Which the Patient Safety Evidence Report Identifies a Need for More-Scientific Evidence*

Name of Detailed Table in Evidence Report	Number of Health Information Technology Projects Addressing the Practices		
	Planning	Implementation	Value
Evidence Report: Impact and Effectiveness			
Greatest strength of evidence	0	1	0
High strength of evidence	1	3	0
Medium strength of evidence	12	15	10
Low impact or strength of evidence	1	7	0
Lowest impact or strength of evidence	4	5	1
Evidence Report: Further Research			
Likely to be highly beneficial **	12	17	8
Likely to be beneficial ***	5	8	3
Evidence Report: not rated, but covered in projects	22	17	14
Practice not addressed in evidence report	1	0	1

* Many projects are addressing health IT that falls into more than one category.

** Evidence report lists 30 practices, including, for example, perioperative glucose control to reduce surgical-site infections, changes in nursing staffing, CPOE, use of analgesics for patients with abdominal pain, and use of bar coding for patient identification.

*** Evidence report lists 29 practices, including, for example, use of hip protectors for falls and injuries, use of crew resource management (teamwork), antibiotic-impregnated catheters, and geriatric consultation services for hospital-acquired complications.

CONCEPTUAL FRAMEWORK FOR IMPLEMENTATION OF HEALTH IT

To guide our collection of information on the experiences of the health IT grantees, we constructed a conceptual framework that identifies the key components involved in ensuring effective adoption of health IT by health care organizations. This framework builds on the work of Greenhalgh et al. (2004), as well as two systematic reviews on health IT conducted by RAND. We used the framework to develop sets of questions to discuss in interviews with the AHRQ-funded health IT grantees, to ensure that we gathered information on the key components of this framework (see the Technical Appendix for methods). Findings from those interviews are summarized in the next section of this chapter.

We conceptualize health IT development in any given organization as having three phases: *adoption*, *implementation*, and *sustainability* (Figure 4.1). At each phase, the *external environment* and the *internal organizational commitment and support* affect decisions about what technology to implement, the scope and timing of implementation, and the likelihood of sustainability.

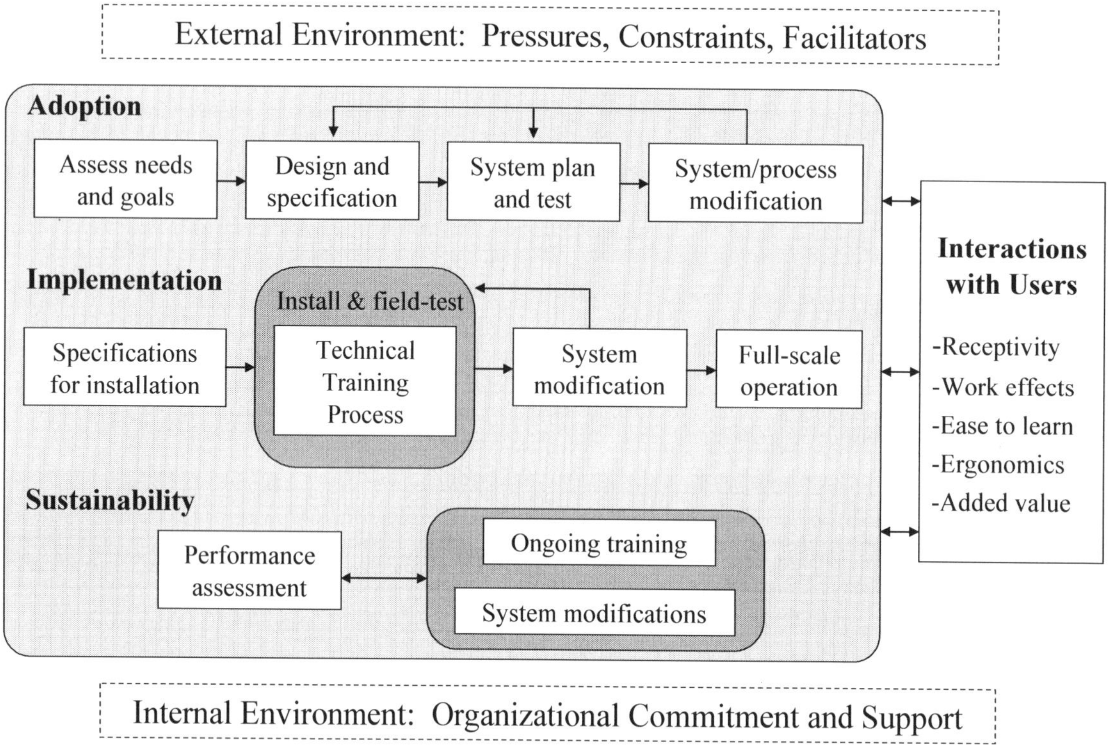

Figure 4.1 Conceptual Framework for Health Information Technology

The *adoption* process begins (or should begin) with an extensive planning process involving the assessment of the needs and goals of an organization, system, or community for use of health IT. It leads to the selection of a product and vendor, aids in the development of a plan and test for the new health IT, integrates the health IT into existing workflow processes and systems, and modifies the system as required to achieve maximum benefits from the technology. *Implementation* of health IT requires detailed specifications that are most often developed by technology specialists within an organization and by the health IT vendor. Once designed and built, the health IT system is installed during a field-test period, and end users are trained on its use. After the workflow changes are made to fully integrate the health IT into the health care setting, the system is moved to the level of full-scale operation. *Sustaining* a health IT solution requires continuous effort and investment, particularly with regard to training of new and existing staff to maintain familiarity with the technology and its enhancements.

The involvement of end users in each phase is critical for their ownership of the technology. Factors that influence end users' acceptance include their ability and willingness to learn how to use the health IT, purchasing or designing health IT that is easy to use, and, above all, implementing health IT that adds value to the individual user. Users must be willing to try new technology, but the technology also has to make their job easier.

EXPERIENCES OF THE HEALTH IT GRANTS

Having identified key components associated with the adoption, implementation, and sustainability of health IT innovations in the conceptual framework, we developed three semi-structured interview protocols (i.e., one for each of the planning, implementation, and value grants) that addressed the components of this framework. The protocols also included questions to seek feedback to AHRQ on the grant process, as well as support from AHRQ, the Health IT Resource Center, and the Coordinating Center. Grantees were provided with a copy of the interview questions in advance. Refer to the Technical Appendix for details on the methods and the interview protocol.

We conducted one-hour interviews with the planning grantees between mid-May and early July 2005 (approximately three-quarters of the way through their planning grant). Of the 38 grantees, 34 participated in the interviews (the four "missing" grantees were contacted five times but either did not respond or the interview was scheduled and ultimately cancelled). We conducted telephone interviews with approximately half of the implementation grantees between June and September 2005 (21 of the 40 grantees). These interviews were completed with all the grantees by the end of October 2005. These interviews occurred approximately eight months to one year into their three-year grant period. We conducted semi-structured interviews with half of the value grantees (13 of 26 projects) in August and early September 2005. The interviews occurred approximately ten months to one year into the three-year period of these grants.

Planning Grantees

All of the health IT projects were led by partnerships of organizations, and the planning grant PIs reported that most of their partnerships had the full support of their organizations' leadership. They also had working teams made up of a wide range of staff. In many cases, a business relationship already existed between the partners before the PIs were awarded the AHRQ funding, and their roles were well-defined, usually with one lead organization having responsibility for managing the project. In other cases, the partners were brought together because they shared a common patient population or health care priorities.

As described above, the planning grants included a number of projects that were wholly or partly rural (Table 4.3). For the rural projects, the rural partners were usually direct providers of health care services; and most such projects, rural and urban partners were participants in their business relationships. Any partner organizations located in urban areas generally had the greater resources in terms of finances, IT expertise, IT capabilities, and research experience. Many of the partnerships in rural projects did not have formal arrangements before the planning grant, but they had shared common patient populations. Planning-grant funding provided the mechanism to develop more-formal relationships to address an identified need.

The planning grantees reported that their decisions to implement health IT were primarily internally motivated, rather than driven by external pressures. A number of grantees reported that they were planning to implement multifunction health IT interventions. A number of the projects were focused on getting relevant patient information into the hands of providers at or before the point of care, regardless of where the patient sought care.

Lessons learned. Grantees reported encountering three broad categories of challenges during the planning-grant cycle: managing the partnership and process, finding sustainable resources, and initiating or maintaining end-user involvement. In addition, grantees identified challenges to be handled at the start of the implementation cycle, including persuading organizations to make the necessary institutional investments in health IT and resolving remaining technical and legal issues.

The grantees found that user resistance was the largest and most obvious barrier to making process changes. In most cases, end users were involved in some way in planning for health IT implementation. Most often, they were involved in needs assessments, the results of which guided the planning team on what type of health IT to implement and provided input on the type of information that should be available from the health IT that is implemented.

The AHRQ planning grants had a number of positive effects on these local projects. Because of this funding, these communities were able to bring partners to the table and to establish a formal planning process for the adoption of health IT. Grantees felt that the funding provided legitimacy for the projects, which encouraged other stakeholders to pledge their support. In several cases, interviewees said that the planning process for health IT would never have reached a high enough priority level (with the many competing investment and staff needs) without the AHRQ funding. In addition, having AHRQ oversight for the project provided a focus for developing specific goals for the partnership and for making the project a priority for the leadership of the organizations involved, as well as for the team members. This focus and enthusiasm are imperative to sustaining a multi-organizational project and creating a firm foundation for a multiyear endeavor. In addition, many grantees have also been able to leverage the funding to identify other potential resources for implementation. When asked if the grantees could proceed with implementation without AHRQ funding, the group was split roughly down the middle. Some thought that they would need further AHRQ funding; others thought that they might be able to find other funding, but that their projects might have to be altered significantly to continue (e.g., a slower pace, a narrower scope of work, fewer partners, and fewer lives affected).

Replicability. Despite these challenges, nearly all of the grantees concluded that other organizations could replicate the planning process. The key components of success are standard in any large change process: collaboration, effective leadership, and trust. Specifically, many grantees identified that

- the planning process takes much longer than they anticipated and can be much more difficult than they thought.
- because of changing priorities or situations, projects should allow for flexibility in design and should use an iterative planning process, even if it takes longer and costs more money.
- setting realistic expectations with leadership and within the partnership is important to accomplish early in the project.

Some organizations indicated that multi-institution partnerships may be easier to replicate in rural rather than urban communities because the institutions tend to have many existing interdependencies. This view was balanced by reports that the rural partners often had difficulty making significant contributions to the projects because of limited resources—staff, finances, and expertise. Organizations that partnered with such "low-resource" organizations expressed concern about the ability of those organizations to sustain participation over time.

The grantees tended to agree that the planning and development process is different for each community, as determined by unique local needs and externalities. They reiterated that technology alone is not enough and that projects will need to address the "human side of things," including personal interactions, relationships, and building of trust among each partners and around the technology. (This conclusion is supported by an extensive literature on the social aspects of implementing technology, which is reflected in the conceptual model for implementing health IT presented in Figure 4.1.)

Implementation Grantees

Lessons Learned. Grantees identified a variety of lessons from the first eight months of implementing their three-year projects. These lessons included practical issues (such as the need to identify and obtain IT support within their organizations, and managing vendors to do what they contract to do), as well as more-complex issues of communication and cooperation within and across organizations. They found that the change process needs to be managed proactively, including effective communication, especially when working across multiple decisionmakers. They also observed that it takes more time than expected to do every task, particularly when an evaluation accompanies implementation. In view of these lessons learned, the implementation grantees identified the following success factors:

- A sense among participants that health IT has value or is inevitable, which contributes to greater willingness by employers and payers to reward implementation of health IT.
- Having an open process that encourages participation by end users, ultimately leading to buy-in as people get enthused when they see the potential for the health IT system to give them more control over what they do, and to get better results.
- The need for top management to have a clear focus on and visible commitment to the health IT, as expressed by expectations for results and persistent support during implementation.
- Provision by top management of adequate financial resources and staff who have the needed expertise and readiness to carry out the implementation process.
- Knowledge within the organization that comes from prior experience with health IT implementation, which enhances the ability to create products that are easy for physicians and other clinical staff to use, especially if the new health IT complements and integrates with other products.
- Competition between sites, which stimulates greater motivation and perseverance to complete the health IT implementation and achieve high-quality IT products.

Grantees reported experiencing a number of challenges, which they had anticipated, and which they thought others also were likely to face. Some of these were operational issues, such as the need to transition information from large numbers of existing forms into bar codes and other standardized formats, and short-term negative effects of health IT on productivity, which end users are learning to work with. Others related to end users' reactions to the new systems, especially physicians whose "buy-in" is essential for successful implementation. In particular, middle-aged and older physicians are not as willing as younger ones to adopt, so it is a challenge to get them to use the system. Community physicians also are especially concerned about legal and confidentiality issues. Additional issues identified include conflicting project priorities and competing resource commitments within the organizations; complex and substantial legal barriers; difficulty in keeping up with rapidly changing technology (hardware, software); and staff stress as a consequence of complex and rapid changes with introduction of new systems.

The implementation grantees also experienced some issues that they did *not* anticipate, such as significant limitations in IT technical "know-how" within their organizations, the need to change vendors and external consultants to get appropriate support, the interdependency across sites for consistency in software and procedures, and the impact of leadership changes on the project. The extent of end user expectations also were surprising, including the demand for one-on-one training with physicians and the need for a longer duration of training, as well as the need for continued modification and refinement of application software for EHR and CPOE, to meet end users' requirements.

Our early review of the proposed evaluation designs and measures of the implementation grants suggested that grantees had varying skills in evaluation work, and many of their evaluation designs were weak. Some implementation grantees reported that they included professional evaluators as part of their project teams, and we found that their evaluation designs tended to be more robust. In addition, some of the grantees reported that they were tracking process or impact measures outside the scope of their AHRQ funding. These grantees stated that this work is not part of the AHRQ funding, so it is possible that they will not be reporting the results of those measures to AHRQ at the end of their projects.

Replicability. When asked whether other organizations or communities could replicate what they are doing, most of the grantees we interviewed said they could. Some were working to create explicit models that could be replicated. A number of grantees mentioned that other organizations would need to have either IT expertise and substantial funding or partners with such resources, and that their success would depend on the organizations' sophistication. The majority of those we interviewed said that they had to adapt current "off-the-shelf" technology to fit the needs of their specific application, which others likely would have to do as well.

Value Grantees

The purpose of the value grants is to assess the value of health IT, with each grant assessing a specific aspect of IT. At the time of our interviews, most of the value-grant teams were just starting health IT implementation or have just completed baseline data collection for their evaluations. Their early experiences offered some preliminary insights. Many of these experiences related to barriers and facilitators to the implementation of health IT, and they tended to mirror those emerging from the planning and implementation projects. All of the grantees reported that other organizations and communities could replicate their projects, if they had the resources and a similar environment. In terms of evaluating the value of health IT, grantees noted the following lessons:

- Few validated instruments are available for measuring the value of health IT interventions.
- Defining new data elements is important and challenging; the grantees need peer input—both internal and external—to explore how to define and operationalize data elements and analyze the data.
- New analytic methods need to be developed through multidisciplinary approaches in order to attain a better understanding of the complexity of the processes involved—for example, patient-provider communication.
- Literature relevant to rural hospitals is limited.
- Obtaining reliable data on costs is difficult.
- Sampling is complicated because the population that serves as the sampling frame often turns out to be smaller than had been expected; in some cases, the sample may be too small to provide the statistical power needed for inference.

- For larger evaluations, time lines can change because there are multiple moving parts (e.g., changes in partners, concurrent actions) over time.
- It is best to introduce the work to providers' practices as a quality-improvement effort, rather than as research; if providers think it is research, providers expect it to end.

ISSUES AND ACTION OPPORTUNITIES

The existing patient safety grants continue to produce a fast-growing literature on patient safety epidemiology, which can help decisionmakers assess future patient safety priorities. AHRQ-funded projects should continue to expand the epidemiological knowledge base as results of newly awarded projects are added to those of the earlier projects.

The health IT projects are among those recently funded projects. As with AHRQ's other patient safety projects, they address a diversity of patient safety issues and practice settings. They also are contributing new evidence about the effectiveness of a variety of patient safety practices for which the patient safety evidence report has identified that additional evidence was needed. According to reports from the health IT grantees, the funding that AHRQ has provided is supporting development work they could not have carried out without these additional resources. Thus, it appears that AHRQ is making a significant contribution to the adoption of health IT across a range of health care practice settings. The partnerships across communities developed in these projects, and lessons learned by the grantees in working with those partnerships, can serve as guidance for others to enhance their adoption of health IT.

Issues to Consider

Not surprisingly, the early experiences of the health IT projects revealed many of the same design and implementation issues identified by previous patient safety projects. They also raised numerous other issues related specifically to use of IT systems, which are summarized in this chapter. This information is valuable for other organizations embarking on similar projects. It also can help the National Resource Center on Health Information Technology to fine-tune its technical support for the grantees.

Suggestions for AHRQ Action

- **AHRQ should maintain an ongoing monitoring process that uses data from the national network of patient safety databases, as well as published research, to examine shifts in trends for patient safety epidemiology in specific aspects of health care and to identify emerging safety issues that need to be addressed to ensure the safety of health care practices.**

Much of the recent patient safety work has focused on priorities identified in earlier published research and reporting processes, based on documented incidence rates of patient safety events. Recent patient safety efforts may have started to change these incidence patterns, with possible improvements made for earlier priorities even as new issues may be surfacing. An ongoing analysis of patient safety trends should draw upon available data in the published literature, existing reporting systems, Census data, and other national or state data sources. Benchmarks should be created by establishing baseline counts or rates for specific patient safety issues, with analyses of changes in incidence or rates in subsequent years. When a national patient safety data repository is established, the data in the repository could serve as a primary source of data for regular analyses of trends. However, given that a national repository will be based on voluntary reporting, inconsistencies in reporting behaviors could lead to fluctuations in reported events that do not actually reflect underlying changes in patient safety conditions. Therefore, estimates of events or rates will need to be triangulated, using multiple available data sources, including published epidemiological papers.

- **AHRQ and the Resource Center should establish a structured program of start-up support and training to first-time grantees to help them understand their responsibilities and AHRQ's expectations.**

New health IT grantees need training on both the grants-management process and the technical aspects of the projects' work. Many of the current grantees do not have previous experience working with federal grants, and they find the process confusing and time-consuming. The Resource Center was not in place in time to support the early work for the first round of health IT grants, but it is now well positioned to provide support for future grantees. The grantees identified three areas of assistance: (1) Provide early feedback on the appropriateness of each project's scope; (2) provide guidelines or criteria to grantees to assist decisions on whether to use in-house staff or an outside vendor to develop the IT technology, and for selecting effective and responsive vendors; and (3) provide faster and easier access to information on what all grantees are doing, to support all grantees with mutual problemsolving.

- **AHRQ should encourage and support work on development of flexible, inexpensive, IT solutions that are accessible for low-resource and rural organizations.**

A major barrier for organizations to invest in health IT is their lack of resources in finances, personnel, and IT expertise, coupled with the fear that the minimal investments they are able to make will be wasted when technology changes. Helping to identify health IT solutions that are flexible and affordable for organizations with varying degrees of IT capabilities, yet that can retain interoperability capabilities, would be very useful for these low-resource organizations, as well as for partnerships with a mix of highly sophisticated IT and entirely paper-based systems. An important corollary is for AHRQ to assist low-resource organizations with identifying what health IT solutions will be most beneficial for them. Organizations with low patient volumes and highly limited referral paths may require far less in the way of health IT solutions than more-complex organizations.

- **AHRQ should explore mechanisms to strengthen the evaluation component of the health IT implementation projects, such as providing more evaluation technical assistance, requiring that grantees partner with evaluation researchers; or having the evaluation component be conducted independently from the health IT implementation.**

The weak evaluation designs for the implementation grantees included a significant number of grantees with no control groups and, in some cases, use of only post-intervention measures. As a consequence, there may not be valid data for assessing the effects of the implementation work, and the results will not be generalizable to other settings. Several of the grantees signaled that they were having difficulty trying to understand which outcomes to evaluate and how to construct appropriate measures. Attention also should be given to measurement methods needed for effective evaluation and to assessment of costs and cost-effectiveness (for which we had a recommendation in *Evaluation Report II*; see the Appendix).

- **AHRQ should work with the Resource Center to clarify to grantees the functions of the Resource Center and the type of technical assistance it can provide to grantees,**

and to tailor technical assistance on evaluation measures and methodologies to the unique features of local projects.

The health IT grantees varied widely in their knowledge of the role of the Resource Center and how best to use this resource. These findings virtually replicate earlier feedback from the previous patient safety grantees about the Coordinating Center. It is an ongoing challenge for a support function of this type to communicate to its clients what it can do for them and how. With this early feedback from the health IT grantees, a focus can be placed on addressing the issues identified.

- **In partnership with other relevant federal agencies, AHRQ should develop clear federal guidance on standards and other requirements for interoperability of health IT, making the investment in health IT more compelling and easier for low-resource organizations.**

AHRQ is well positioned to support the recommendations of other federal agencies regarding health IT standards and to influence the adoption of these standards through their grant process. Such support and influence would facilitate the spread of interoperable technology and would be consistent with standards established for a national patient safety data repository.

CHAPTER 5. PROCESS: BUILDING INFRASTRUCTURE FOR EFFECTIVE PRACTICES

> ***Building Infrastructure for Effective Practices:*** *Establishment of the health care structural and environmental elements needed for successful implementation of effective patient safety practices, including an organization's commitment and readiness to improve patient safety (e.g., culture, information systems), hazards to safety created by the organization's structure itself (e.g., physical configurations, procedural requirements), and effects of the macro-environment on the organization's ability to act (e.g., legal and payment issues).*

This chapter focuses on the fourth of five system components of an effective patient safety system, as depicted in Figure 1.1. This component is building infrastructure for effective practices, as defined in the box above.

BUILDING FROM *EVALUATION REPORTS I* AND *II*

As noted in our previous reports, building a supportive infrastructure is critical for successful adoption of improved patient safety practices throughout the United States. In *Evaluation Report I*, we delineated a number of infrastructure elements, including patient safety culture, information systems, adverse-event-reporting systems, interdisciplinary teams, multi-institutional collaborations, and quality-improvement systems and measures. We also highlighted how the FY 2000–FY 2001 patient safety projects were addressing infrastructure issues, and we explored the contributions of other AHRQ-funded vehicles to patient safety infrastructure development.

Infrastructure building takes place at national (e.g., laws and regulations), regional (e.g., state reporting systems), and local (e.g., safety culture, teamwork) levels. In most cases, the work carried out in the patient safety initiative is contributing to infrastructure by providing knowledge, skills, or tools that people at the regional or local levels can use in their infrastructure-building activities. For example, a patient safety culture survey can be used by hospitals to assess their cultures and take actions to strengthen them, or provision of training can help build skills for people responsible for implementing state reporting systems.

In each year of the evaluation, we focus on key aspects of the patient safety initiative that have the potential to contribute to infrastructure development in the field, with updates provided in later years on the progress made on many of them. These topics are chosen according to information obtained in annual interviews conducted with AHRQ leadership staff, as well as from our assessments of the relative importance and effects of the various activities being undertaken. In *Evaluation Report II*, we focused on three aspects of AHRQ's infrastructure development activities: establishing partnerships with other organizations, the Patient Safety Improvement Corps, and consumer involvement in patient safety. We presented baseline data for FY 2004 on the extent to which AHRQ and other national-level organizations were engaged in partnerships for patient safety strategies and action. This analysis will be repeated in FY 2006 to identify changes in partnership patterns as the patient safety initiative moves forward. Similarly, the progress of the PSIC has been tracked each year.

Our evaluation of infrastructure development draws upon information provided by a variety of data sources, including relevant written materials and documents, and interviews with AHRQ staff, other federal agency staff, participants in AHRQ-sponsored programs, and PIs for AHRQ-funded grants or contracts. Our assessment addresses the following key questions:

- To what extent is there an infrastructure of interorganization partnerships that is pursuing collaborative approaches to improving patient safety practices?
- How are AHRQ activities and funded projects contributing to the establishment of infrastructure to support national implementation of effective patient safety practices?
- What additional research or development work is needed to strengthen the establishment of effective infrastructures for patient safety practices in the health care system?

In *Evaluation Report III*, we address other aspects of AHRQ's continued efforts to establish and strengthen partnerships and working collaborations, with a focus on several new AHRQ initiatives for creating formal collaborations to reinforce adoption of practices, each of which is discussed here. We also present updated findings from FY 2005 for our evaluation of the PSIC training program.

HIGH-RELIABILITY ORGANIZATIONS

In 2004, the Knowledge Transfer (KT) program replaced the User Liaison Program (ULP), which had long been part of AHRQ's outreach strategy for dissemination of knowledge and tools from ARHQ-sponsored research. The KT program builds upon the strengths of the ULP, and its special emphasis is on collaboration with end users in the field. It has two important new features: (1) Inclusion of hospitals, purchasers, and insurers among policymakers and stakeholders of interest; (2) a focus on selected prime topics of concentration, with strategies to help end users improve their practices. The program is based on the "learning network" model, in which participants with shared interests come together voluntarily to increase knowledge by sharing ideas, setting standards, and building tools to strengthen the work of the member organizations (Wenger, McDermott, and Snyder, 2002; Snyder and Briggs, 2003; Bessant, Kaplinsky, and Morris, 2003).

AHRQ has identified and begun work on four key learning-network initiatives: high reliability organizations for patient safety, care management in Medicaid, purchaser/provider interactions, and pediatric asthma. For each initiative, AHRQ has contracted with an external organization to serve as the lead for carrying out the work, with one AHRQ KT staff member collaborating and providing policy guidance. Relevant AHRQ staff are also actively involved in the development and implementation of the initiatives. Our evaluation focuses on the high-reliability organization initiative, because it is the one that addresses patient safety specifically. A general definition of a high-reliability organization is one that operates effectively and efficiently, with minimum error. One of the goals of this initiative is to delineate the specific features or components of a high-reliability health care organization, and to determine the implications of this definition for the participating hospitals.

The initiative on high-reliability organizations is working with seven to 12 health systems to maximize their ability to interact effectively among themselves, using the learning-network model. The health systems that were selected to participate in the initiative are innovators in patient safety. As of September 2005, they had begun to identify their learning needs, including coaching on how to implement new patient practices, information systems to support the work, use of leadership rounds on inpatient units, and other leadership roles. AHRQ is developing a strategy to respond to these needs, which will draw upon the products and information generated by the patient safety grantees and other related AHRQ projects. The challenge will be to find the common areas in which users' needs match the available products, as well as to adapt existing products or develop new ones to best respond to those needs.

HOSPITAL SURVEY ON PATIENT SAFETY CULTURE

Early in the patient safety initiative, the QuIC sponsored development of the Hospital Survey on Patient Safety Culture, funded by AHRQ, which was intended to help hospitals assess how well their cultures emphasize patient safety, facilitate open discussion of errors, encourage error reporting, and create an atmosphere of continuous learning and improvement. The survey contains 51 questions that assess 12 areas, or dimensions, of patient safety. It was developed by Westat using a process that included a review of existing research on safety, error and accidents, and error reporting; an examination of existing published and unpublished safety-culture-assessment tools; and interviews with hospital employees and administrators to identify key patient-safety and error-reporting issues. The survey was pilot-tested in a sample of 20 diverse hospitals across the United States; more than 1,400 completed surveys were received from staff in those hospitals (AHRQ, 2005b).

Upon receiving Office of Management and Budget approval in early 2005, AHRQ, in partnership with the American Hospital Association (AHA), began to disseminate the hospital culture survey to U.S. hospitals and health care systems. A coordinated set of informational, training, and support activities was undertaken in this process. Three technical-assistance conferences calls were held in February through March 2005, and more than 400 hospitals participated (AHRQ, 2005c). A growing number of hospitals and QIOs across the country are using the culture survey to take early action for improving identified patient safety issues. In addition, hospitals have been using several other patient safety culture surveys that are available.

To support further use of patient-safety-culture surveys in health care, AHRQ is establishing a contract for continued technical support for use of the hospital survey, development of new culture surveys for other types of providers, and establishment of a benchmarking database for culture-survey results by the end of 2007. AHRQ plans to establish a family of Surveys of Patient Safety Culture by adding questionnaires for physicians, long-term care facilities, and ambulatory care.

We note that the administration of a culture survey is just a first step in the process of achieving a strong patient safety culture. The hospitals (or other organizations) using such a survey need to apply the information it generates to guiding actions for changing practices and attitudes, to ultimately achieve a strong safety culture, which is the ultimate measure of success for this aspect of safety infrastructure. However, as of the end of FY 2005, it was too early to have information on how proactively hospitals have used the survey to bring about such change.

THE PATIENT SAFETY IMPROVEMENT CORPS

The PSIC is a partnership program between AHRQ and the Department of Veterans' Affairs (VA) National Center for Patient Safety (NCPS). AHRQ leads the initiative and is providing $7 million over four years to train teams from every state in the country; the VA organizes and conducts the training sessions. The primary goal of the PSIC is to improve patient safety throughout the nation by increasing the number and capacity of health care professionals with core patient-safety knowledge and skills that can be applied directly.

The annual PSIC training program consists of three one-week sessions in September, January, and May. The training is composed of didactic sessions led by NCPS, AHRQ, and other experts; homework and reading assignments to complete between sessions; and a team patient safety project. Participants progress from learning basic patient-safety principles and concepts in the first session to training in more sophisticated skills, such as statistical techniques for assessing patient risks, in the second session. In the third session, each state team presents its patient safety project and results. The sessions focus on the practical application of patient safety science, change implementation and management, medical-errors reporting and analysis, medical/legal issues, and patient safety tools. In addition, graduates of the training had access to the VA staff, PSIC instructors, and the members of their or other PSIC teams, as support resources, as well as the PSIC training materials and the AHRQ or VA Web sites.

Eligible participants are teams of state staff in the field and staff from up to two of each state's selected hospital partners (for a total of four participants maximum per state). The PSIC program is tuition-free, and teams selected to participate are reimbursed for airfare, lodging, per diem, and local travel costs. Each participant also receives books and other resource materials.

In 2003–2004, the first year of the PSIC, 15 state teams completed the program; in 2004–2005, 21 state teams completed the program. State teams accepted to the 2005–2006 PSIC program were notified in August 2005. Teams from all 50 states were expected to be trained by the end of the third year, after which the program may be expanded to a train-the-trainer model in order to broaden its reach to more individuals and organizations.

The RAND evaluation of the PSIC program is designed (1) to provide feedback to AHRQ and the VA on the participants' experience with the program and suggestions for improvement; and (2) to assess the extent to which the knowledge and skills gained from the PSIC training have been used by the participants to improve patient safety. The information is based on group interviews conducted with 11 of the 15 state teams participating in the first PSIC round at their final training session in May 2004, 12 of the 21 states participating in the second PSIC round in May 2005, as well as individual one-year, follow-up telephone interviews with 38 graduates of the first PSIC program about ten months after completion. Follow-up interviews with the second group will be conducted in spring 2006.

Team Composition and Formation

As required by AHRQ, each state team comprised representatives from both the state government and hospitals located in the state. In 2003–2004, state participants tended to be employed by state health departments in a regulatory capacity, which was similar to the first-year trainees. Compared with the first-year trainees from hospitals, more of the second-year hospital trainees tended to be the patient safety officers or to hold other positions with responsibilities directly related to patient safety, perhaps reflecting increased national awareness of the importance of patient safety. There was also a noticeable increase in the number of team members who were affiliated with QIOs. In 2003–2004, team membership remained stable over the course of the year-long training. In the following year, seven of the 12 teams interviewed reported changes in membership or that some members had to miss some sessions.

Experiences Reported by PSIC Trainees

Overall, the short- and longer-term observations of the first two groups of PSIC trainees were very positive. Participants valued the tools and skills they learned and were continuing to use many of them in their positions day-to-day. They appreciated and continued to draw upon the technical aspects of the training; the hands-on exercises, especially the knowledge gained through their own and other teams' projects; and the extensive reference materials and library provided as part of the program. Additionally, they continued to view as significant resources the course's networking opportunities and the broader perspective they gained about patient safety. Significantly, there were strong indications that the PSIC program in both years had contributed to actions in the field to improve patient safety, and that it was an important force in building a national infrastructure to support implementation of effective patient safety practices.

During the 2003–2004 training year, many state and hospital representatives shared information and materials with colleagues back home, and they were pushing to implement patient safety initiatives in a variety of areas, many directly related to their PSIC team project. One year later, these PSIC graduates had used many of the PSIC skills and tools to make meaningful changes on a variety of patient safety fronts, as shown in the interview responses summarized in Tables 5.1 (state participants) and 5.2 (hospital participants).

Use of the PSIC Training to Achieve Practice Changes

States have used information gained through the PSIC training to initiate or influence legislation, modify adverse-event-oversight procedures, and improve existing state reporting systems. Hospitals also said that the PSIC training was an important factor in modifications they have made to adverse-event-oversight procedures. Additionally, hospitals have used the skills and tools learned through the PSIC to promote a patient safety culture within their organizations and share data across organizations in an effort to better understand causes of error.

Table 5.1
How the PSIC Training Influenced Patient Safety Actions by States, Reported in One-Year Follow-Up Interviews with the 2003–2004 Trainees, 2005

Patient Safety Action	Percentage Responding "yes" (*n*=15)
Initiation of or influence on regulation(s)/ legislation	47%
Modification of hospital oversight procedures when an adverse event occurs (e.g., change content of Root Cause Analysis/RCA)	47*
Modification of an existing state reporting system to improve how it captures patient safety issues or how information is reported to others	33
Creation of a statewide reporting system	20
New membership in or formation of a patient safety coalition of stakeholders	20

* For 7 percent of the respondents, this question was not applicable, or not relevant to the respondent's type of organization or role within that organization, or the respondent could not answer the question.

Table 5.2
How the PSIC Training Influenced Patient Safety Actions by Hospitals, Reported in One-Year Follow-Up Interviews with 2003–2004 Trainees, 2005

Patient Safety Action	Percentage Responding "yes" (*n*=23)
Modification of processes to review/analyze adverse events or errors	83%*
Promotion of patient safety culture	78*
Sharing data across organizations to better understand causes of error	52
Other changes in review of adverse events	48
Other state- or organizationwide initiatives	48*
New membership in or formation of a patient safety group of stakeholders	35
Creation of institutional adverse-event-reporting system	30

* For 4 percent of the respondents, this question was not applicable, or not relevant to the respondent's type of organization or role within that organization, or the respondent could not answer the question.

Similarly, the 2004–2005 PSIC graduates have mastered a set of skills and have been sharing the skills and tools learned in the training with others in their organizations, as well as more broadly in their local communities and across their states. They have drawn on these resources to launch new patient safety initiatives and improve existing ones. Notably, there was an early awareness among the 2004–2005 trainees of the necessity for diverse parties to collaborate (e.g., hospital employees and state regulators). Part of this change from the previous year may be attributable to the increased interest in and awareness of patient safety issues nationally, and the ensuing realization by these parties of the benefits of collaboration. According to the attendees, the PSIC has played an instrumental role in changing attitudes. The experiences of the first (2003–2004) group, coupled with the national trend of increasing awareness of patient safety issues, seemed to have paved the way for somewhat easier interactions in the 2004–2005 group.

Trainees noted some barriers that created challenges for their participation in the program and their ability to make changes at home, ranging from lack of resources (e.g., time, funds) to cultural obstacles. The trainees also underscored a need for continued training beyond the end of the third PSIC session, and they voiced the need to train larger, more-diverse teams that include front-line clinicians, high-level decisionmakers (e.g., chief executive officers [CEOs]), and senior staff from both hospitals and states.

Many PSIC graduates expressed interest in becoming a trainer by participating in PSIC train-the-trainer sessions. They also identified the resources they would need to meaningfully assume the role of a trainer, including outside funding for time, travel, and teaching supplies; educational materials from the VA and AHRQ; reduced work responsibilities; support staff to handle administrative details; and a refresher course. In addition, it will be important for trainers to have access to off-the-shelf tools that can be put to work readily in their training activities.

TWO NEW AHRQ NETWORKS

AHRQ has been funding a number of program initiatives that serve as vehicles for transferring knowledge, conducting applied research, and engaging end users in the health care system in adopting proven safety practices. These programs include the Integrated Delivery System Research Network (IDSRNs), the Centers for Education and Research on Therapeutics (CERT), Practice-Based Research Networks (PBRN), User Liaison Program (ULP), Translating Research Into Practice (TRIP), and Partnerships for Quality (PFQ). Most of these initiatives have operated with limited budgets and small staffs. In *Evaluation Report II*, we assessed the use of these networks for enhancing safe practices, concluding that they had the potential to play substantial roles (Farley et al., 2007).

We learned in interviews with AHRQ staff that AHRQ intended to extend these networks as a means of accelerating dissemination of proven patient safety practices for adoption by health care providers, which we reinforced in a recommendation that encouraged such a strategy. During FY 2005, AHRQ indeed moved along these lines by initiating two innovative networks—"Accelerating Change and Transforming Organizations and Networks" (ACTION), which replaces the former IDSRN program, and "Developing Evidence to Inform Decisions about Effectiveness" (DEcIDE), which adds a new direction for its dissemination activities (AHRQ, 2005d; AHRQ, 2005f).

We identify ACTION and DEcIDE in this chapter, because AHRQ's establishment of these new programs has expanded the infrastructure needed to support strategies to accelerate adoption of safer health care practices. In Chapter 6, we discuss the contributions these programs are making to AHRQ's overall dissemination strategy to support practice adoption. ACTION is supported not only by AHRQ but also by a number of other federal agencies and private organizations. The consortia that were awarded the first ACTION contracts consist of large and diverse sets of partnering organizations. AHRQ plans to focus the work of these networks on projects that move proven practices into the field, scaling up implementation and innovation from the local delivery system to regional and national levels. DEcIDE is a collaborative research and practice-based network program that emphasizes development and dissemination of new evidence and knowledge about effectiveness, costs, and safety of various therapeutic approaches. These networks will assist AHRQ and other federal agencies with implementation of the Medicare Modernization Act of 2003.

ISSUES AND ACTION OPPORTUNITIES

As described at the start of this chapter, our evaluation identified patient safety culture, information systems, adverse-event-reporting systems, interdisciplinary teams, multi-institutional collaborations, and quality-improvement systems and measures as key components for a patient safety infrastructure. Over the first three evaluation cycles, our assessment of ARHQ's contribution to each of these components has found that, thus far, ARHQ has made its strongest contributions to building patient safety culture, information systems, adverse-event-reporting systems, and multi-institutional collaborations. It has done so through the types of patient safety projects it has funded, its partnership outreach with other organizations, its specific initiative for high-reliability organizations, dissemination of the Hospital Survey on Patient Safety Culture, the new ACTION and DEcIDE networks, and other network programs. In addition, our analyses have found that the PSIC has stimulated actions by its participants to implement patient safety improvements in their respective organizations.

Issues to Consider

In considering what more needs to be done to ensure the establishment of effective infrastructures for patient safety practices in the health care system, it will be important to assess not only AHRQ's contributions but also those of other organizations that have been undertaking similar actions at the national, regional, and local levels. AHRQ's patient safety initiative has not yet generated tangible products relating to interdisciplinary teams and quality-improvement systems and measures. However, AHRQ has been working on products for health care providers to strengthen interdisciplinary teams, which it plans to introduce in FY 2006, and other AHRQ programs are supporting work to strengthen quality-improvement systems and measures. In addition to new products, continued work is needed for already-established programs.

The Hospital Survey on Patient Safety Culture, which offers hospitals across the country an effective tool for assessing their status and pursuing data-based actions to build effective patient safety cultures in their facilities, could have far-reaching effects on performance if it is widely used. It is critical for AHRQ to capture information on the use of the survey and resulting improvement activities by hospitals, so that the effect of the survey can be documented and assessed.

The first two years of the PSIC have demonstrated the value of bringing together groups of diverse stakeholders and of providing them with intense patient safety training. With completion of the training for a third group of participants in May 2006, the original goal and scope of the PSIC will have been achieved. However, the very success of this PSIC program raises questions about how to establish a sustainable infrastructure for building on the expertise and partnerships it has fostered thus far. AHRQ and the VA face some important decisions regarding how to approach the next generation of PSIC activities, including options for a train-the-trainer program, continued support of patient safety activities by PSIC graduates, and strategies to reach health care decisionmakers, who ultimately determine how far these activities can proceed.

Within the framework of the collective work under way by various organizations, AHRQ will need to choose strategically where to invest its limited resources. Engaging in partnerships is a useful strategy to leverage resources and avoid duplication of effort, but AHRQ will need to decide which types of partnerships might be the most fruitful to lead, and which are better led by other organizations. The new AHRQ network initiatives for high-reliability organizations, ACTION, and DEcIDE offer rich potential to contribute to the patient safety infrastructure. We encourage AHRQ to monitor these programs regularly, seeking feedback from participants, and gathering information to learn from experience and modify the programs as needed.

Suggestions for AHRQ Action

- **As AHRQ continues to work with hospitals to support their use of the Hospital Survey on Patient Safety Culture, it should establish a structured monitoring process that collects data on trends in hospital use of the survey, issues identified by hospitals from survey data, and actions hospitals take to respond to those issues.**

AHRQ has worked proactively to move the culture survey into the field, and it is establishing a contract to provide continued support for its use. Through this support function, AHRQ has an opportunity to document systematically the diffusion of use of the culture survey across providers, its effects on providers, and actions to achieve safer health care. This information can help guide AHRQ's work, in collaboration with other organizations, to make available tools, technical assistance and other resources to help hospitals' safety-improvement efforts (see Chapter 6).

- **Building upon the successful PSIC training that has reached the front-line hospital and state-level staff, AHRQ should work with relevant organizations to stimulate outreach to increase commitment to patient safety by key decisionmakers who are needed to make patient safety improvements happen.**

The intensive training model used for the PSIC to train front-line hospital and state-level staff has been very effective. However, to make meaningful and lasting changes, front-line staff need the support of key decisionmakers—the individuals with authority to make policy and resource decisions (e.g., senior management, state legislators). AHRQ should consider ways to gear a modified PSIC program to the busy schedules of these decisionmakers, to engage them in recognizing and addressing patient safety issues so that individuals at all levels of the health care system are moving toward similar goals. For example, a Web-based or one-day in-person short course could offer executives the knowledge to help them make informed decisions about patient safety policy and practices.

- **AHRQ should assist PSIC graduates in finding support for their continued engagement in patient safety issues, updating their skills and knowledge, and encouraging synergy among the PSIC graduates and with others in the field.**

AHRQ should continue to support its original investment in training for the core group of PSIC graduates so that initial gains in patient-safety knowledge and skills are retained and further improvements can be made. AHRQ might offer periodic "refresher" courses in flexible formats (e.g., Web-based training, telephone conference calls, in-person short courses), and it also could provide ongoing access to technical assistance and opportunities for interactions among the PSIC graduates. This follow-up support should adapt to the changing needs of the PSIC graduates over time.

- **If AHRQ and the VA pursue a PSIC train-the-trainer program, it should be a working partnership among AHRQ, the VA, and the trainers to ensure that the persons who become trainers have the needed teaching skills to fulfill the defined trainer role.**

The continued diffusion of the skills and tools taught through the PSIC is critical to its long-term success and impact. A train-the-trainer program offers the potential to encourage this diffusion by leveraging AHRQ resources into the field through the trainers. The training activities have two potential audiences: (1) staff at each trainer's home institution and (2) staff in other organizations that have not received any PSIC training (e.g., ambulatory clinics, state medical societies, long-term care providers, insurers, accrediting agencies). However, resources will be required to support the trainers' work that will involve a long-term commitment to support the trainers and, to some extent, the people they train. To ensure the success of this effort, it will be important to develop a feasible strategy before embarking on the training process.

CHAPTER 6. PROCESS: ACHIEVING BROADER ADOPTION OF EFFECTIVE PRACTICES

> ***Achieving Broader Adoption of Effective Practices:*** *The adoption, implementation, and institutionalization of improved patient safety practices to achieve sustainable improvement in patient safety performance across the health care system.*

This chapter focuses on the last of five system components of an effective patient safety system, as depicted in Figure 1.1. This component is achieving broader adoption of effective practices, as defined in the box above.

BUILDING FROM *EVALUATION REPORTS I* AND *II*

AHRQ's current challenge in achieving broader adoption of effective patient safety practices is to ensure that the wealth of information emerging from the projects it has funded gets into the hands of end users, along with appropriate tools and assistance to facilitate their efforts. In *Evaluation Report I*, we established baseline information on dissemination activities and published output from patient safety grantees, and we also outlined a suggested approach that AHRQ could use to develop and carry out a cohesive dissemination strategy.

We extended this work in *Evaluation Report II*, again monitoring output from the patient safety grantees, and we examined existing AHRQ funding vehicles and activities that could support dissemination and adoption of safety practices (e.g., IDSRNs, PBRNs). We also reported early information showing that the grantees in the various groups of patient safety projects were reporting similar lessons on what is required for successful implementation of new patient safety practices.

Finally, we examined AHRQ's efforts to define a strategy for dissemination and adoption of patient safety practices, noting that an effective dissemination strategy differs substantially from AHRQ's traditional public-relations activities. We found that, as of September 2004, AHRQ had established a limited structure for dissemination, and that it did not yet have a written strategic dissemination plan. We also stated that AHRQ would need to engage end users to become more familiar with the types of information sources they rely on and to learn how best to package the information so that it is responsive to their needs and interests.

For this report, we continued monitoring AHRQ's dissemination activities. To frame our assessment, we used a conceptual model of the role of a change agency—in this case, AHRQ—in the dissemination and diffusion of innovations. We also heightened our focus on actions taken, recognizing that the AHRQ-funded projects had begun generating awaited results and that a cohesive dissemination strategy was needed to most effectively get those results into the hands of users. The key questions addressed in this chapter are

- To what extent is new evidence on effective practices and implementation methods being disseminated to the broader health care system?
- What actions has AHRQ undertaken to prepare for disseminating the information and tools from the FY 2000–FY 2001 grants and contracts in patient safety?

- How are AHRQ's dissemination strategy and activities contributing to the growing number of patient safety implementation initiatives being initiated by others?

To address these questions, the RAND evaluation team drew upon information from published documents, interviews with patient safety grantees, interviews conducted with AHRQ project officers and leads of funded projects, and discussions with individuals from AHRQ, the Coordinating Center, and the Patient Safety Grantee Steering Committee.

FRAMEWORK FOR ACHIEVING ADOPTION OF EFFECTIVE PRACTICES

As AHRQ embarks on its dissemination activities to stimulate adoption of patient safety practices by U.S. health care providers, it is important that its approach and strategy be grounded in existing evidence of what it takes to accomplish diffusion of a new practice (or other innovation). Such an approach will help ensure that AHRQ gain the greatest possible impact for the resources it commits to this process. Rogers' well-tested model of diffusion of innovations is a rich source of this knowledge (Rogers, 2003). According to this model, AHRQ is functioning as a change agency in carrying out its dissemination activities. A *change agency* is an organization that seeks to influence decisions by end users to adopt or reject an innovation. In this section, we present a framework that defines a change agency's functions according to the Rogers diffusion model. This framework can be used by AHRQ to guide its future dissemination plan and strategy, and it also serves as a useful evaluation tool for assessing the current status of AHRQ's dissemination work.

According to diffusion theory, the diffusion of an innovation (in this case, new patient safety practices or tools) will be influenced by the extent of proactive interventions to encourage its adoption, which may range from a totally passive approach to a carefully planned and managed intervention for change (Rogers, 2003; Greenhalgh et al., 2004). Diffusion models identify the end users as the ultimate decisionmakers regarding adoption of an innovation. Other important players are the developers and commercializers of the innovations, as well as change agencies. Adoption rates usually are slow initially; for successful innovations, rates of adoption increase more rapidly after a critical mass of the end-user population (~10 to 20 percent) has adopted. The portion of the curve between 10- and 20-percent adoption is the important phase of the process, because, after that point, further diffusion will continue spontaneously (Rogers, 2003).

Two-Step Dissemination Model

A two-step strategy emerges from these diffusion models that AHRQ, as a change agency, can use to increase the diffusion and adoption of innovations (Rogers, 2003; Valente and Davis, 1999). This model is shown in Figure 6.1. *Dissemination* and *diffusion* refer, respectively, to the active and passive spread of an idea. Under the two-step dissemination model, change agencies proactively disseminate innovations to carefully selected opinion leaders, who are well-known and respected members of a business or professional group. The opinion leaders then diffuse the ideas to other potential adopters via interpersonal relationships. *Adoption* represents the decision to accept an innovation, and *implementation* entails putting the innovation into practice. Importantly, adoption and implementation are not inevitable consequences of diffusion. Rejection, reinvention, and discontinuance of an innovation can be rational decisions and occur frequently. Input and feedback represent the importance of the end users' perspectives on an innovation (Rogers, 2003; Greenhalgh et al., 2004).

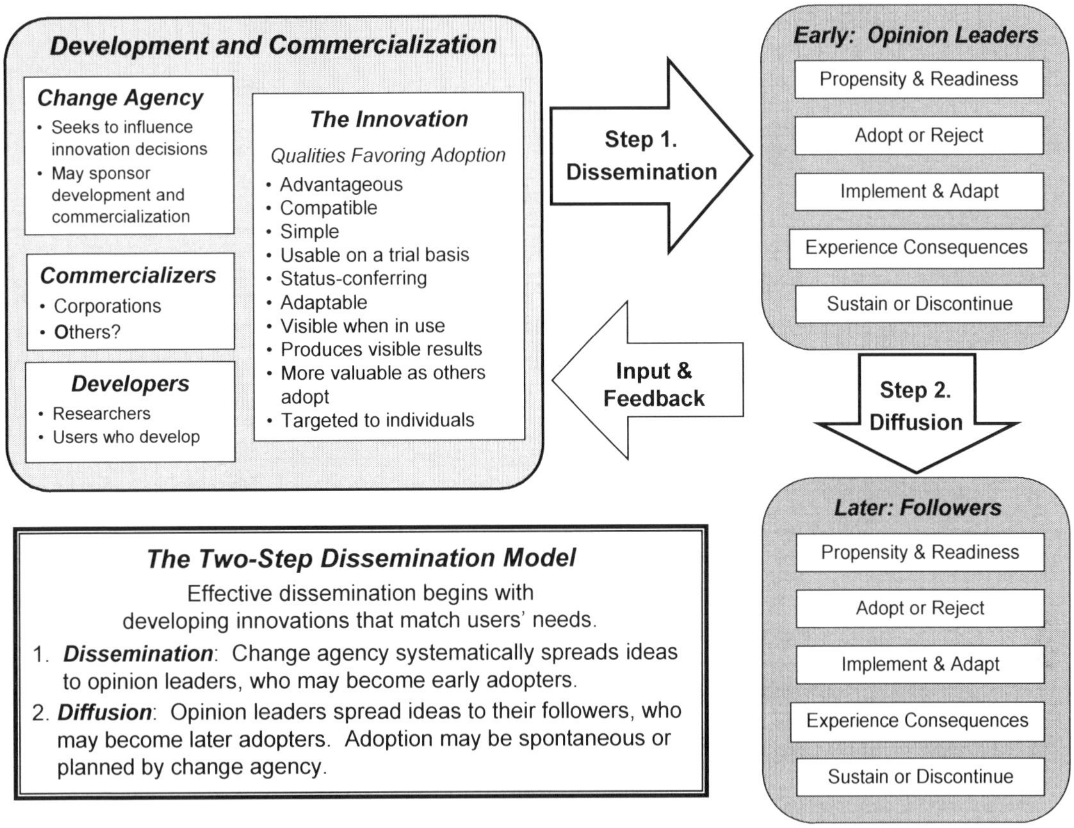

Figure 6.1 A Two-Step Model for Dissemination of Innovation

To the extent that these perspectives are understood when an innovation is being designed, the innovation is more likely to be adopted widely by the target audience. If not fully captured early in the innovation process, this input can be provided through user feedback as the innovation is being used in the field. The model's strength lies in the fact that scientific information alone does not persuade most people, including health care providers, to adopt innovations. Rather, most people rely upon the subjective experiences of previous adopters whom they know and respect (Rogers, 2003; Valente and Davis, 1999).

Model Components

The *development and commercialization phase* is when an innovation is being identified, developed, tested, and packaged for introduction to end users in the field. *Adoption* or *rejection* of an innovation is influenced by the attributes of the innovation, as perceived by the target audience, rather than those envisioned by developers or demonstrated scientifically. Adoption is facilitated when an innovation is advantageous relative to alternative ideas, compatible with potential adopters' values and experiences, simple, usable on a trial basis, status-conferring, visible when in use, and productive of visible results (Moore and Benbasat, 1991). Adaptability, meaning the ability to modify an innovation, greatly enhances adoption and promotes sustainability (Rogers, 2003; Greenhalgh et al., 2004). Change agencies, developers, and commercializers all play roles in the development and commercialization process, their goal being to design an innovation that has value for the end users.

Potential adopters vary widely in their propensity to adopt innovations in general and in their readiness to adopt a particular innovation. *Propensity to adopt* is positively associated with financial resources, social connectedness, educational achievement, scientific literacy, and risk tolerance, among other qualities. *Readiness* is having a perceived need for a particular innovation (Rogers, 2003; Greenhalgh et al., 2004).

Figure 6.1 emphasizes opinion leaders and followers because they represent the majority of potential adopters and are able to make diffusion a self-sustaining process. In the United States, opinion leaders are typically earlier adopters. Most other people are followers, meaning that they adopt later and prefer to obtain innovation information from local opinion leaders rather than from mass media or other outside sources.

The decisionmaking steps of potential adopters entail acquiring knowledge, developing a favorable or unfavorable opinion, and then deciding to adopt or reject. Individuals and organizations frequently adapt innovations to suit their needs and may alter their practices to incorporate the innovation. These change processes are central to adoption, because an organization needs to "unlearn" some of its current practice before it can effectively adopt new innovations. Implementing innovations consistently creates both positive and negative consequences, as well as direct and indirect ones. Ultimately, positive consequences lead to confirmation of an adoption decision, whereas negative consequences produce disenchantment.

Change-Agency Actions to Encourage Adoption

In addition to the innovation-diffusion model, AHRQ can be guided by the experiences of a dissemination model operated by the federal government—the Cooperative Extension Service of the U.S. Department of Agriculture (USDA). The mission of the Extension Service is to translate research findings on new farming technology into practice by farmers across the country. A critical lesson from its experience is that any research program that has the goal of effecting changes in practices must establish *research utilization* as the foundation that guides its organization and action priorities. Several related factors influenced the success of the Cooperative Extension program, which are encompassed in the following two principles (Rogers, Eveland, and Bean, 1976):

- Research is more likely to be utilized if it *solves end users' problems and addresses their needs.*
- The *internal organization of the change agency should emphasize research utilization.*

Working with the components of the two-step dissemination model, along with these two principles on research utilization, we developed a list of actions that should be undertaken by a change agency throughout its product research, development, and dissemination process. These actions are summarized in Table 6.1, along with a summary of the current status of AHRQ's patient safety initiative and related dissemination activities for each of these actions.

Table 6.1 shows that AHRQ has done extensive work in the problem-recognition and -assessment phase of the dissemination process, including the growing body of work by funded projects for testing implementation of patient safety practices. For the remaining phases—development and packaging of innovations and the dissemination and diffusion of those innovations—AHRQ activities have been undertaken on a relatively small scale. The absence of a comprehensive dissemination plan, as well as the continuing unavailability of results and products from the patient safety projects, has continued to hamper AHRQ's ability to move forward with a focused dissemination strategy.

Table 6.1
Change-Agency Actions Taken by AHRQ to Support Diffusion of Patient Safety Practices and Products

Change Agency Action by Innovation Step	Actions Taken by AHRQ
Problem recognition and assessment	
• Sponsor research to identify problems	Funded extensive patient safety research.
• Seek end-user input in identifying problems	Input through Patient Safety Summits; no direct user input into awarding or execution of grants.
• Sponsor basic and applied research on identified problems	Funded extensive research on new practices; increasing emphasis on applied work.
Development of usable innovations	
• Work with developers to ensure innovations have qualities that match users' needs and preferences	Not much work in this area yet due to incomplete information on project results.
Packaging of innovations	
• Recognize that lead users sometimes develop their own innovations	No actions yet to identify, encourage, or coordinate with lead users.
• Focus on higher-priority innovations first	Culture survey and teamwork identified as priorities; none yet identified from projects.
• Package innovation to facilitate adoption	Packaging of culture survey and teamwork; nothing yet from projects to package.
• Prepare and test tailored messages for specific audiences (audience segmentation)	No actions yet on this strategy.
• Develop incentives to encourage adoption	Quality Connect project gives funding; no other direct actions on this strategy.
Dissemination and diffusion of innovations	
• Ensure adequate resources for dissemination	Some resources allocated; need to match resources to dissemination plan.
• Use mass media to create awareness of innovations	Activities undertaken on general safety issues and specific projects; other priorities not set.
• Identify appropriate opinion leaders	Starting to do this with the Steering Committee and dissemination network.
• Pursue direct-contact strategies to persuade opinion leaders to adopt	Direct action—PSIC, Quality Connect project;* also partner in SCIP, 5 Million Lives
• Pursue strategies to promote diffusion from opinion leaders to their followers	No action yet on this strategy.

* Quality Connect is a project to help states take action on quality performance findings in the AHRQ quality and disparities reports (see later discussion in this Chapter).

PRODUCTS GENERATED FROM PATIENT SAFETY GRANTEES

In July 2005, we updated information presented in *Evaluation Reports I* and *II* on publications and other products generated by the patient safety grantees, and added information for the 109 new health IT grants. The types of project-related products identified include journal articles, AHRQ publications, conference presentations, editorials, reporting systems, bibliographies, book chapters, newspaper articles, software products, taxonomies, and government reports. As in past years, we searched library databases and the Coordinating Center Web site to identify products from the portfolio of 99 AHRQ, HRSA-funded, and challenge-grant patient safety projects. We also identified articles published in the four-volume compendium, *Advances in Patient Safety: From Research to Implementation* (AHRQ, 2005a), which are not yet listed in the standard health literature databases.

Table 6.2 presents the results of this review. We found a total of 1,677 documents related to patient safety authored by patient safety grantees during the total time period covered, of which 276 documents were deemed to be related to the AHRQ projects. These project-related documents represent 16.5 percent of the total patient safety–related documents identified, increasing from 8.9 percent last year. The number of products emerging from the patient safety portfolio has continued to increase over time. A total of 139 products was produced from July 2004 to June 2005, doubling the total number of products compared with the cumulative production for all preceding years. These included 90 journal articles, 39 AHRQ publications, three editorials, two conference presentations, and one each of a book chapter, reporting system, survey, and government report.

Table 6.2
Number of Written Products Produced by Patient Safety Grantees, 1997–2005*

Year	Project Related	Not Project Related	Not Clear if Related	Total
Year Unknown	3	0	4	7
1997	0	67	0	67
1998	0	84	3	87
1999	0	95	1	96
2000	0	130	8	138
2001	4	144	15	163
2002	40	124	67	231
2003	68	210	96	374
2004	79	193	66	338
2005 (half year)	73	83	11	167
Under review **	9	0	0	9
Total	276	1,130	271	1,677

* Many of the leads of AHRQ-funded projects are working on other patient safety projects. Therefore, some of their patient safety publications are not from the AHRQ-funded projects, as distinguished in the columns "project-related" and "not project-related."

** Articles sent to journals but not yet accepted, which were listed on the Coordinating Center Web site.

Note: Some of the counts of articles differ from last year's because counts were added for articles that were found on the Coordinating Center Web site from previous years for which PIs were not listed as authors.

DISSEMINATION ACTIVITIES FOR FY 2000–FY 2001 GRANTEE PRODUCTS

With a growing number of published products emerging from the AHRQ-funded patient safety projects, the information needed from these projects to update the scientific evidence and disseminate products to end users is increasingly available. In *Evaluation Reports I* and *II*, we identified and reinforced the need for AHRQ to have a written plan for synthesizing and disseminating the products emerging from the patient safety grantees. In this report, we again focus on AHRQ's dissemination strategy, this time in the context of increasing availability of project results. We also examine specific aspects of a dissemination strategy, working with the two-step dissemination framework introduced above.

AHRQ Dissemination Strategy

The dissemination activities for the patient safety initiative are being carried out by an AHRQ team that cross-cuts the agency, including CQuIPS, the Office of Communications and Knowledge Transfer (OCKT), and the Coordinating Center. CQuIPS is leading activities for publication of research results from the patient safety projects in compendia or journal supplements, and it is guiding product-packaging efforts. The Coordinating Center is working extensively with the individual grantees to identify their written results and other products, and, ultimately, to play a role in synthesizing those results. OCKT is leading the communications process to disseminate the project results to the end users and other stakeholders. As of September 2005, AHRQ had a draft dissemination plan to guide its work. The plan is being updated regularly as the work is refined. With future updates, the diffusion model presented in this chapter would be a strong foundation to ensure that AHRQ is taking a comprehensive and strategic approach to its dissemination activities.

At this time, syntheses of full sets of patient safety project results have not yet been performed, so the only dissemination actions that have been taken are for information and products generated by individual projects. As in previous years, the agency continues to have difficulty getting information from the grantees on what they are generating, either because grantees want to get results published in peer-review journals before disseminating them via other vehicles or because some project officers have limited communication with grantees and do not know what products grantees are generating. In response to this issue, the Coordinating Center was tasked with establishing a central database containing comprehensive information on each patient safety project. A similar database was being built for the health IT grants, and plans are to integrate the two databases to establish one central information resource.

The Coordinating Center started developing this database during 2005, drawing on information from proposals, quarterly reports, and direct communications with each of the grantees. The information needed by the OCKT to tell the stories from the projects is typically not found in higher-level reporting or descriptive information about the projects. With the Coordinating Center responsible for pulling together the information, the database offers the potential for resolving the information problem, which will enable the remaining dissemination activities to move forward with greater ease. We will continue to track this activity to assess the extent to which it is successful in overcoming the difficulties in obtaining information from the grantees on their project findings and products.

Synthesis of Research and Product Development

Before synthesis activities can be pursued, the results from the patient safety projects must be published in the health care literature and then be aggregated for consideration as a group. In addition to the large increase in peer-reviewed journal articles published in the past year, AHRQ and the Department of Defense (DoD) achieved a milestone in 2005 with the publication of the four-volume compendium *Advances in Patient Safety: From Research to Implementation* (AHRQ, 2005a). Another volume of the *Advances* is planned to publish findings from the working-conditions projects. In addition, a journal supplement was being prepared that will publish papers from the Safety and Quality by Design conference held in 2003. Using this approach, AHRQ can compile all the information on a given patient safety issue in one source, making it easier to synthesize and assess evidence for new practices.

Three mechanisms have been identified for performing syntheses of patient safety project results: the Coordinating Center; the KT staff in the OCKT; and a new contract awarded in September 2005 for a Clinical Decisions and Communications Science Center. The new Center is intended to help AHRQ implement provisions of the Medicare Modernization Act of 2003 on the comparative effectiveness of clinical therapeutics and appropriate clinical approaches, and also to assist in promoting the usability and comprehension of complex scientific information for health care decisionmakers (AHRQ, 2005e). The work on patient safety issues to be performed by these resources has yet to be determined.

An important "next step" after collecting and synthesizing the project results is to update the patient safety evidence report to incorporate this new evidence. AHRQ has not yet decided when a formal update of the evidence report will be done, and issues remain unresolved regarding what is acceptable evidence for patient safety practices.

In the first half of 2005, NQF appointed a panel to update its patient safety practices, having decided to proceed in the absence of an evidence-report update. The standard of evidence being used is "preponderance of evidence," rather than scoring the level of evidence. In addition to assessing effects on patient outcomes, the panel expanded the conceptual frame to consider impacts on intermediate processes (e.g., impact of simulation on length of learning curve). The report from this process is intended to update the previous evidence report.

Packaging of Products and Tools

In this discussion, we make a distinction between packaging and commercializing products based on the results of AHRQ-funded patient safety projects. The work that AHRQ is doing is referred to as packaging products, in which generic products are developed to make it as easy as possible for health care providers to use them and adapt them to their particular circumstances. Commercialized products, on the other hand, are developed by firms in the private sector with the intent to market them for sale to health care providers. Both packaging functions are useful aspects of getting research results into practice; we address here AHRQ's product-packaging activities.

The packaging of products and tools for practices identified as important from the patient safety grantees' work must await completion of the synthesis of project results and identification of these practices. However, using individual project results, AHRQ has acted on other opportunities for product and tool development, as well as on other work it has funded as part of the patient safety initiative. For example, OCKT produced some tools, checklists for care processes, and a number of products targeted to consumers (e.g., staff surveys, training packages).

Some grantees have started working with commercial firms to commercialize products emerging from their work. AHRQ staff believe that a great deal of the product-development work will happen in the field, as firms see profit opportunities in developing and selling the products. As shown in our dissemination framework (Figure 6.1), such commercializers are important players in the diffusion of innovations. In choosing priorities for product development, AHRQ can make a strong contribution by focusing on products that are important to end users but that are not likely to be picked up by commercial firms.

Working with its Steering Committee, the Coordinating Center is developing a dissemination network of provider associations, information groups, and other organizations with patient safety interests. These network contacts are examples of the opinion leaders identified in the two-step dissemination framework. The Coordinating Center plans to work with them to reach others in the field for diffusion of information and practices. These network members also help guide priorities for product development and testing.

Two major products that AHRQ has developed directly are the hospital culture survey and a package on training and implementation for improved teamwork. AHRQ has collaborated with the DoD on the teamwork training products, which were planned to be available for dissemination by the end of 2005.

In the RFA for the PIPS, AHRQ placed an emphasis on packaging project results for dissemination. The17 newly funded PIPS grants are expected to develop tools from their work, including a comprehensive implementation toolkit that focuses on "...documenting the impact of the safe practice intervention on patient care and the manner in which barriers to implementation and adoption were overcome" (AHRQ, 2004c). In this toolkit, it would be helpful to provide a model of innovation adoption at the health care provider level, similar to the model presented in Figure 6.1, which could be supported by evidence from the organizational-change literature.

Communication for Dissemination

In the communications portion of the dissemination process, OCKT is coordinating work within the agency, stimulating cross-center development work and decisionmaking. OCKT is working with AHRQ staff to develop a shared understanding of a market-oriented approach. OCKT has done some focus groups with stakeholders to get information on their needs. They also have established partnerships with other organizations, one example being collaborations with the AHA and the American Medical Association on the Five Steps Toward Safety for consumers. One priority for AHRQ is the need to respond to congressional requests for documentation that the patient safety initiative is achieving improvements in care. To do this, AHRQ staff tend to cite results of specific successful projects.

Some of the patient safety dissemination vehicles currently being used are the AHRQ publication *Patient Safety Findings and Actions*, program briefs on individual projects, and communication of patient safety results in speeches, testimony, press releases, and other communications activities. AHRQ has developed a campaign to put safety ads in the Washington, D.C., metros, and it has worked with the Advertising Council to bring patient safety messages to the public through its partnerships with others.

Electronic media—in particular Web sites—are an integral part of any dissemination strategy. AHRQ introduced the new Web site Patient Safety Net (PSNet) in FY 2005, which is intended to be the national "go-to" place for information on patient safety. PSNet is one among several Web sites addressing patient safety. Other Web sites include patient safety information posted on the main AHRQ Web site, QualityTools, and the PSRCC Web site for patient safety grantees. In addition, Web sites of outside groups (i.e., Premier, IHI, the Joint Commission, the National Patient Safety Foundation [NPSF]) are other important patient safety resources. Users often have to check all of the Web sites to find the information they are seeking, and AHRQ is finding it difficult to coordinate the posting of information among its sites. Transparent linkages among the Web sites could help alleviate this issue, with the goal of creating a seamless resource.

In August 2005, AHRQ announced a new initiative called the *Quality Connect* project, which is designed to help states take action on quality performance findings in the AHRQ quality and disparities reports (AHRQ, 2004b; AHRQ, 2004a). Quality Connect has been funded with $1 million for technical assistance to states to work toward improvement on measures in the quality or disparities reports, which include some patient safety measures. The program also has provisions for a robust communication and action infrastructure, sharing of successful practices, and a national information resource. It offers a useful example of a structured dissemination strategy that uses a partnership approach to support implementers in performance-improvement efforts.

TWO NEW AHRQ NETWORKS

As discussed in Chapter 5, AHRQ initiated ACTION and DEcIDE in FY 2005 as innovative networks for practice testing and dissemination (AHRQ, 2005d; AHRQ, 2005f). The establishment of these new network programs creates additional infrastructure intended to accelerate the translation of existing or emerging evidence about patient safety into practice. Furthermore, with a diversity of participants in each network, the participating networks are positioned well for collaborative work among providers, researchers, insurers, and others, including AHRQ, to enhance diffusion of research results into practice in the field.

The activities undertaken in these two new programs contribute to AHRQ's dissemination strategy for improving quality of care, including patient safety as an important component of that strategy. We summarize the programs here and document briefly AHRQ's approach and expectations for working with them. We note that AHRQ already has established other similar programs, such as the Centers for Education and Research on Therapeutics, Practice-Based Research Networks, and EPCs. The two new programs are "next-generation" models that extend beyond existing programs to push AHRQ's collaborative strategy further toward active dissemination of new practices for adoption in the field.

ACTION, which is the successor to the IDSRN program, will retain successful features of the IDSRN program, such as its focus on projects that are "demand-driven" by those who manage, deliver, or receive health care services; rapid-turnaround projects; and a competitive bidding structure using task orders within a master contract. It differs from the IDSRN by providing for inclusion of a much more diverse and larger set of organizations in the ACTION partnerships or consortia (i.e., QIOs, governmental organizations, employers or employer coalitions, and others that partner with provider levels to promote evidence-based health care improvements). In addition, significantly more funding will be targeted toward projects that scale up implementation and innovation from the local delivery system to regional and national levels. To help fund these efforts, AHRQ has established a number of interagency agreements and commitments from a wide array of other federal agencies (e.g., CMS) and private foundations (e.g., Robert Wood Johnson Foundation).

DEcIDE is a collaborative research and practice-based network program that will assist AHRQ and other federal agencies with implementation of part of the Medicare Modernization Act of 2003. Its primary objectives are to rapidly develop scientific evidence and new analytic tools to assist health care providers, patients, and policymakers in making informed decisions about the comparative effectiveness, appropriateness, and outcomes of health care services, particularly prescription medications. In contrast to ACTION, the DEcIDE program places more emphasis on developing new evidence and knowledge about effectiveness, costs, and safety of various therapeutic approaches, including pharmacoeconomics, pharmaco-epidemiology, and clinical and health services research using increasingly sophisticated data warehouses and databases maintained by health care organizations.

AHRQ has strongly emphasized that it expects consortia participating in ACTION and DEcIDE to collaborate with investigators or use evidence from other AHRQ initiatives and programs, such as the Centers for Education and Research on Therapeutics, Practice-Based Research Networks, and EPCs. AHRQ also emphasizes the importance of the willingness of partners in different consortia to work together on some projects (AHRQ, 2005d; AHRQ, 2005f). These developments appear to reflect a recognition by AHRQ that achieving substantial changes in patient safety and other quality areas will require the following:

- Meaningful involvement of a wide array of stakeholders
- Rapid and flexible funding mechanisms that reduce traditional barriers to practical research and facilitate sharing
- Adoption of new knowledge and innovations across a broad range of health care organizations.

The Clinical Decisions and Communications Science Center (see section above on AHRQ's dissemination activities) is intended to translate scientific evidence developed by these network programs into targeted products for patients, providers, policymakers, and the public. This center also is to facilitate greater access to and understanding of scientific evidence, provide insights into the various factors that influence health care decisions, and identify partners for disseminating and leveraging information in collaboration with AHRQ (AHRQ, 2005d).

CAN A PATIENT SAFETY COOPERATIVE EXTENSION MODEL WORK?

As AHRQ has prepared for the dissemination and adoption of improved safety practices, it has been seeking mechanisms in the field to establish direct and sustainable working relationships with end users. Such direct relationships are essential to maintaining momentum in providers' understanding and ownership of patient safety issues and solutions, which, in turn, will drive the diffusion of effective practices through the health care system.

The USDA Cooperative Extension Service has been identified as a model that might be adapted to the health care system (Cooperative State Research, Education, and Extension Service Web site). This program has the mission of advancing agriculture knowledge in the field by supporting research, education, and extension programs that bring that knowledge to the users. The program has an extensive network of state, regional, and county extension offices across the United States. that serve as the front line for dissemination of new technologies and practices.

An existing network in the health care field that is similar to the USDA Extension model is the state-level QIOs, with which CMS has contracted to ensure that Medicare beneficiaries receive appropriate and timely care. The QIOs have recently been charged with educating the public about information on provider quality (CMS, 2005; Sprague, 2002), and as discussed in the next section, they are an integral part of several national-level quality-improvement initiatives in collaboration with providers.

In an assessment of the Extension Service, Rogers and colleagues identified three factors that influenced the success of the program, which can be applied equally to health care (Rogers, Eveland, and Bean, 1976):

- Research is more likely to be utilized if it solves end users' problems and addresses their needs. In health care, both providers and patients are the end users of research, and patient safety research is most likely to be utilized when it addresses needs of both patients and providers.
- The internal organization of the extension agency should emphasize research utilization and be supported by adequate funding. The organizational culture, promotion criteria, and internal incentives should reward efforts to increase research utilization.
- The extension service should have a carefully planned strategy for bringing about research utilization, including a critical mass of new technology to be disseminated and a relatively homogeneous and well-defined group of potential users. The extension staff should have a single vision and a common language, and individuals in the program should have specialized and well-defined roles.

The QIOs are well positioned to fulfill the first and second criteria, which in effect already apply to their existing performance-improvement activities. The third criterion, however, poses a more difficult challenge. Given the diversity of the health care system, at both the market and individual-provider levels, it could be difficult to identify well-defined, homogeneous groups of users. Furthermore, successful adoption of patient safety practices that have been standardized at the national level is likely to require adaptation of practices to specific organizational cultures. In addition, to assume a significant extension role, the QIOs would need larger staffing levels to reach a broad range of providers, and workload demands from this program could disrupt their ability to fulfill their other contractual obligations.

Despite these issues, the extension concept is appealing, and we believe it merits further exploration by AHRQ, CMS, and other collaborators. In particular, further analysis would be helpful in specifying the similarities and differences between the two organizational settings.

OTHER INITIATIVES FOR PATIENT SAFETY IMPROVEMENTS

The momentum of patient safety improvement activities in the field has accelerated substantially since September 2004 (one year ago). Our baseline analysis of partnership networks, presented in *Evaluation Report II*, identified a variety of working partnerships among national-level organizations in 2004. It appears that these partnership structures are generating activity that is reaching the end users—health care providers. Three new national-level initiatives with strong potential to improve safety in health care have recently emerged, and are being led by field-based organizations. AHRQ is participating as a partner in all of them. They are excellent examples of how AHRQ can leverage its role by helping to support patient safety improvement activities by others:

- In 2003, CMS and the CDC established the Surgical Care Improvement Project (SCIP), which seeks to reduce the national incidence of surgical complications by 25 percent by the year 2010 (SCIP, 2002). The QIOs are providing support to participating hospitals, and they also are expanding hospital recruitment.
- The 5 Million Lives Campaign, led by IHI, is an initiative to engage U.S. hospitals in a commitment to implement changes to improve patient care and prevent avoidable deaths (IHI, 2005). The national campaign has the goal of saving 100,000 lives in an 18-month period (by June 2006), and again every year thereafter.
- QIO work on transformation of hospitals seeks to provide "the right care for the right person every time," based on adoption of the IOM six aims for quality care (IOM, 2002). The initiative also encompasses a range of providers, including hospitals, nursing home care, and ambulatory care; and it is establishing a broader set of performance measures.

ISSUES AND ACTION OPPORTUNITIES

Issues to Consider

Our evaluation has shown that AHRQ has done extensive work in the problem recognition and assessment phase of the dissemination process (see Table 6.1), but its activities for the remaining phases thus far have been on a relatively small scale. Although AHRQ has undertaken a number of creative patient safety dissemination activities in this past year, it has made slow progress in both the development and packaging of innovations and the dissemination and diffusion of those innovations. A positive step has been the establishment of the Clinical Decisions and Communications Science Center, which is to translate scientific evidence developed by the various AHRQ network programs into targeted products for patients, providers, policymakers, and the public. However, it is not clear yet how much of the center's work will be focused on patient safety issues. AHRQ has yet to implement a comprehensive strategy for dissemination of patient safety practices and tools, owing to several factors, including difficulties in getting complete information from grantees on their project products and findings, and a tendency to emphasize communications rather than product packaging and dissemination for end users. In addition, end users have not yet had much involvement in AHRQ's process for establishing priorities for products and tools.

At the same time, initiatives for implementing proven patient safety practices have been mobilizing in the field with increasing speed. Recent national initiatives, such as the SCIP, the 5 Million Lives Campaign, and the hospital transformation work by the QIOs, show that commercialization of proven patient safety products is happening in the field, which may be the best place for that work to be done. Yet, AHRQ will need to be able to move quickly to match the pace of these initiatives and serve as a useful resource to them. This issue is highlighted in feedback from health systems involved in the AHRQ's KT initiative for high-reliability organizations, which suggested that it may be too late because health care organizations are already doing substantial patient safety work.

Given this context of strong and growing activities by end users, AHRQ should focus its efforts where it offers unique capability to make the best contributions to the dissemination of patient safety practices. While AHRQ is viewed by end users as the leader in patient safety research and knowledge, the agency is not an organization on the front line of health care delivery—which is where changes in practices need to occur to improve safety. We suggest that foremost among the unique contributions AHRQ can make will be the new knowledge and products from its patient safety projects, including establishment of priorities for patient safety actions based on syntheses of this information. A complementary contribution is service by AHRQ as a national clearinghouse for patient safety information, including results from AHRQ-funded research, products and tools for safe practices, and linkages to information provided by field-based organizations and initiatives.

Suggestions for AHRQ Action

- **AHRQ should develop and implement a strategic plan that defines its focus on and roles for disseminating new patient-safety knowledge and products, with support for this work coming from an appropriate internal infrastructure and budget.**

In *Evaluation Reports I* and *II*, we encouraged AHRQ to develop a cohesive strategy to disseminate the new knowledge and products from the patient safety projects, and this recommendation continues to stand. A draft dissemination plan now has been prepared as an internal working document. This plan should be grounded in the two-step dissemination model presented in Figure 6.1, which we have developed in part to provide this type of structure for AHRQ's use. It also should clearly specify AHRQ's roles relative to those of end users and other implementers in the field, and it should be communicated broadly to those stakeholders to support effective collaboration with them. Broad-scale success in carrying out a dissemination strategy is contingent on the agency's providing it resources appropriate to the amount of work defined in the dissemination plan.

- **As AHRQ continues to partner with health care systems and other implementers on dissemination activities, it should place a priority on synthesizing information and packaging products and tools from the patient safety grantees so that this information can be available to the field in a usable and timely manner.**

Patient safety practices and products vary in their readiness for packaging and dissemination. Practices that already have been documented as proven by the patient safety evidence report, and adopted by the NQF, are the most ready for packaging. Those that are still emerging from the AHRQ-funded patient safety projects are only starting to be evaluated as candidates for best practices. We believe that AHRQ can get the best leverage from its finite resources by focusing its work on the latter group of practices, leaving the packaging of already-proven practices to field-based initiatives. A necessary first step in this process is for AHRQ to amass the collective results from the patient safety projects to serve as the information base for synthesis of the findings. In the synthesis process, end users need to be involved in setting priorities for packaging products for dissemination. To support this product-packaging strategy, it is essential for AHRQ to develop and sustain strategic partnerships with those who are best positioned to do the packaging work and to provide bridges to actual end users.

- **Building on its strength as a national information resource on the scientific basis of patient safety issues and practices, AHRQ should further extend its work to establish the Patient Safety Net (PSNet) as an integrated clearinghouse on patient safety, including linkages to information provided by other organizations, that is the "go-to" place for users across the country.**

Positioning AHRQ visibly as a key national resource takes full advantage of the existing strengths of both AHRQ's research base and the OCKT, which has years of experience communicating with diverse stakeholders and gathering information from them on their needs and preferences. The PSNet was created to become the "go to" place for patient safety information. Although a natural extension of AHRQ's strengths, this vision has not been fully achieved yet because not all the Web-based patient safety information is accessible through the PSNet. Several other AHRQ Web sites also contain information, as do Web sites of other national-level organizations. By establishing one clearly visible entry point, together with a highly navigable Web site, AHRQ can integrate all these sources with seamless links across sites when necessary. The Web site should have clearly designated areas where each key stakeholder group can access the information most important to it (e.g., products and tools for providers, policy products for insurers, published papers for researchers).

CHAPTER 7. PATIENT SAFETY OUTCOMES

A key component of this evaluation is the identification and tracking of measures for assessing the effects of the AHRQ patient safety initiative. We begin this chapter with a reference to the earlier work we have done on this topic, as well as a review of the framework used for the product evaluation (the final component of the CIPP model; see Chapter 1). We follow this review with an assessment of the availability and limitations of data on patient safety effects. We then present baseline trends for selected patient-outcome measures related to safety, including measures for which data already are being reported by various organizations, as well as results of our trend analyses for selected PSIs and other measures developed by patient safety grantees. Finally, we examine issues regarding the sensitivity of claims-based measures to the structure and content of underlying datasets, which can have important effects on estimates of changes in outcomes over time.

BUILDING FROM *EVALUATION REPORT II*

In *Evaluation Report II*, we outlined our approach to this work and documented a variety of data resources that might provide usable information for assessing changes in patient safety outcomes over time. These resources include health care claims data, reporting data on adverse events and medical errors, data from health records, and surveys on patient-safety practices and culture. Currently available resources offer a starting point for assessing trends in patient safety outcomes in the United States. At the same time, it will be important to continue "pushing the envelope" to develop new data sources and measures for health care settings and issues that currently are not being measured.

In this report, our focus for the product evaluation is on the assessment of baseline national trend data for selected patient outcomes related to safety issues. We have defined the *baseline period* to be the late 1990s through 2003, a period that immediately precedes the time that actions undertaken through the AHRQ patient safety initiative should begin to influence patient safety outcomes. Although AHRQ first funded patient safety projects in FY 2000 and FY 2001, the results of these projects, and of other, related activity by AHRQ and its collaborators, would not be expected to have observable nationwide effects until at least two to three years later, and perhaps longer. Encounter data for more recent years will begin to become available by 2006, which can be used to assess whether early effects of patient safety activities become observable in trends for the patient-outcome measures.

FRAMEWORK FOR THE PRODUCT EVALUATION

A detailed presentation of the product-evaluation framework is presented in Chapter 7 of *Evaluation Report II*. We summarize the framework here, to serve as context for the evaluation results presented in this chapter. As discussed in Chapter 1 of this report, the structure of the patient safety evaluation is based on the CIPP model. The fourth component of that model is a product evaluation, in which the consequences of the patient safety initiative for various stakeholder groups are assessed. We are focusing on stakeholder effects that arise from actions taken to strengthen two of the five system components assessed in the process evaluation (see Figure 1.1), with the components of interest being infrastructure development and adoption of effective practices.

The model presented in Figure 7.1 guides our strategy for identifying and analyzing outcome measures. According to this model, actions taken in the health care system for development of infrastructure help support adoption of effective patient safety practices by providers, which in turn should achieve improved outcomes for patients. Both infrastructure development and practice adoption also affect other stakeholders involved in the initiative to create a safer health care system, including providers, states, organizations working on patient safety, and the federal government.

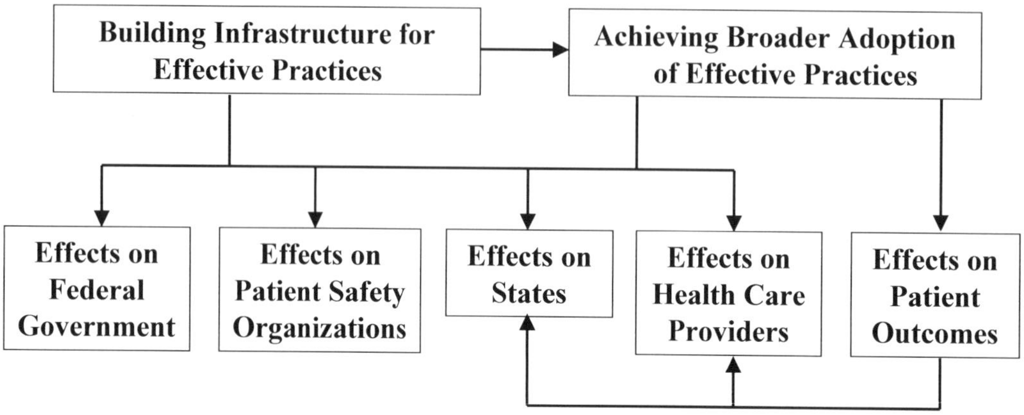

Figure 7.1 Conceptual Model of Potential Effects of the National Patient Safety Initiative

In the remainder of this chapter, we present results of our work to identify a limited number of measures for use in analyses of the effects of the evaluation on patient outcomes, which is the aspect of the product evaluation on which we focused first. (Future work is planned to assess effects on other stakeholders, as well as consideration of practice-adoption progress.) In the next section, the availability and quality of measures are discussed. Then we present baseline trends for selected measures, including some published by other organizations and others that we estimated specifically for this evaluation, using a subset of the PSIs and measures developed by AHRQ-funded patient safety projects. Finally, we summarize the key lessons learned from the estimation and analysis of baseline trends for all the measures identified.

AVAILABILITY AND QUALITY OF MEASURES

Where data on patient outcomes and effects on other stakeholders are available, we have been measuring changes in those outcomes over time. We have focused on three settings—ambulatory care (with primary care as the central component), hospital-based care, and long-term care—recognizing that data will be more readily available for hospital-based issues because most of the patient safety work to date has focused on this setting. The initial analyses have addressed patient outcomes, because what national-level data are available are extremely limited and exist only for outcome measures.

In addition to patient-outcome measures, we have explored data for assessing changes in the process of establishing a strong patient safety infrastructure and implementing safe practices, which are intermediate steps toward improving patient outcomes. In particular, the successful adoption of evidence-based safe practices is an important process outcome that should directly affect patient outcomes.

The absence of national-level data on rates of practice adoption is due, in part, to the lack of well-tested methods for collecting the data. Research studies on specific provider populations have generated some information, finding in general that the uptake of proven safe practices by hospitals and other providers has been slow and requires substantial effort to accomplish (Blake et al., 2006; Leape et al., 2006; McFadden, Stock, and Gowen, 2006). For example, a survey of hospitals in two states found that only 34 percent of hospitals had implemented computerized physician order entry systems for medications (Longo et al., 2005).

Available Patient Outcome Measures and Data Sources

In *Evaluation Report II*, we provided summaries of a number of existing patient-safety data resources. These resources are heavily oriented to inpatient care settings. Furthermore, the existing measures for inpatient care are limited in scope and could be enhanced by adding new measures that address other important aspects of care. The availability of national-level patient safety measures currently is inadequate for long-term care settings and virtually nonexistent for ambulatory care settings, primarily due to the absence of viable data resources. Where data resources do exist to study patient safety, a number of conceptual and technical problems have limited the development of patient safety measures, including challenges in defining what constitutes a measurable patient safety event, issues of incomplete and inconsistent data, and how best to use data for purposes other than the originally intended uses.

Health care encounter records are a key data resource for measuring patient safety outcomes, because they are universally used for the processing of payments for health services provided. In many cases, related data are available through public sources, such as hospital discharge data in the Healthcare Cost Utilization Project (HCUP). Other data resources are less advanced or available, but they are likely to become more important in the future. Examples include state adverse-event-reporting systems, health-related data for long-term care facility patients reported to CMS using the Minimum Data Set (MDS), and the growing availability of EHRs.

Data Quality Issues

Patient safety measures based on encounter data or adverse event reports have some similar problems and biases. Observed variations in values for encounter-based patient safety measures (e.g., PSIs) for different institutions or geographic regions could be due in part to inconsistent coding methods, rather than to real differences in performance. These data-coding problems can be exacerbated by the process of aggregating records from multiple sources into common databases, given that some databases include many codes for diagnoses and procedures, whereas others include few of those codes. The discrepancies mean that encounter-based measures developed for a database such as the HCUP National Inpatient Sample (NIS) may work poorly when applied in other databases with fewer variables.

Deriving nationally consistent patient safety measures from data in adverse-event-reporting systems is also difficult because these systems employ many different reporting standards, making it almost impossible to aggregate the data at higher levels. Practices of state-level reporting systems vary widely, including variations in which events are reported and in how reported data are defined and coded. These systems also are vulnerable to inconsistencies of interpretation, as well as to bias in reporting practices by providers, particularly if the providers are not protected from malpractice exposure for the reported adverse event. The Patient Safety and Quality Improvement Act of 2005 offers the potential to address these issues by establishing consistent definitions and data standards through development of its national network of patient safety databases.

Similar challenges for development of patient safety measures are found for other national-level data sources. For example, MDS data are collected by numerous long-term care providers across the country and are used for evaluating facilities and maintaining federal certification. This regulatory use of the MDS data may lead to bias in providers' reporting of data, which may affect the validity of patient safety measures obtained from this data source. It is challenging to assess the consistency of MDS coding practices across facilities and states, and at least some of the MDS measures are difficult to subject to external validation (Mor, 2004; Mor, Angelelli et al., 2003; Mor, Berg et al., 2003; Roy & Mor, 2005).

Developing New Ambulatory Care Measures

Ambulatory care practice involves consultative and testing services that cross multiple providers and collaboration with patients in managing health care conditions. Therefore, patient safety measures designed for ambulatory care will differ from those used for inpatient care, and they may address care processes more than outcomes. Efforts to develop patient safety measures that cover the spectrum of these services remain at an early stage of development, although some work has been done with medication management and ambulatory surgery (Dovey et al., 2002; Galt et al., 2005; Lapetina and Armstrong, 2002).

Considerable activity and research have been undertaken recently on patient safety in ambulatory care in the development and validation of appropriate taxonomies, reporting systems, best practices, and outcome measures. Recent research on patient safety issues has tended to focus on several categories of events as being important in ambulatory care.

- Communication errors (i.e., events occurring between providers and patients and/or between multiple providers)
- Medication-related errors (i.e., events involving prescribing, transmitting prescriptions, or monitoring medications)
- Failure by clinicians to follow up on diagnostic tests
- Inappropriate use of anesthesia for office-based procedures.

The majority of the ambulatory care reporting systems identified in the literature receive unstructured responses from reporters, which are then coded retrospectively into categories (Bhasale, Miller et al., 1998; Dovey et al., 2002; Lapetina and Armstrong, 2002; Fernald et al., 2004).

Activities by Organizations. Several organizations that historically have focused on patient safety in hospitals are now addressing ambulatory care settings, as are ambulatory care organizations:

- The Joint Commission recently announced its 2006 Ambulatory and Office-Based Surgery National Patient Safety Goals, which include improvements in patient identification, communication among caregivers, medication safety, and health care–associated infection rates; reconciling medications across the continuum of care; and reducing the risk of surgical fires.
- The National Quality Forum modified its list of "30 Practices for Better Health Care" to include practices in ambulatory care settings.

- The National Committee for Quality Assurance added several patient-safety process measures to its Health Plan Employer Data and Information Set, including monitoring of side effects for children prescribed attention deficit–hyperactivity disorder medication, identification of drugs to be avoided for use by the elderly, and annual monitoring of elderly patients on persistent medications.
- The Ambulatory Care Quality Alliance announced a starter set of 26 clinical-performance measures for ambulatory care in May 2005, with the goal to eventually establishing uniform performance-measurement standards addressing both quality and patient safety issues.

Organizations also are starting to address the data needs for patient safety measures and monitoring in ambulatory care. According to the Accreditation Association for Ambulatory Health Care, a growing number of ambulatory facilities are implementing voluntary adverse-event-reporting systems. Currently, a small number of states have mandatory reporting systems for outpatient settings, and even fewer mandate reporting of adverse events that occur in physicians' offices.

AHRQ-Funded Projects. A number of AHRQ-funded research projects have been aimed at developing reporting systems, measurement typologies, and methods to reduce errors in ambulatory care settings. The majority of the projects address medication errors, focusing on outcome measures for specific medications or clinical populations. Several other AHRQ projects have identified potential process and outcome measures for particular aspects of ambulatory care. These studies and measures are still at an early stage of development, and much more work is needed before it will be possible to measure patient safety outcomes in ambulatory care on a national basis. Recognizing this need, AHRQ has given the coordinating center for Practice-Based Research Networks an additional $1 million in funding to test measures and mechanisms for reporting adverse drug events and other events in the participating practices. This work was expected to yield some results by the end of FY 2005, which could offer a foundation for expanding data collections more broadly in the ambulatory-care sector.

BASELINE TRENDS IN PATIENT SAFETY OUTCOMES

In this section, we present baseline estimates for selected patient safety outcomes measures—some drawn from summary statistics reported or published by other organizations and others developed through our own analyses. Those drawn from other sources are the patient safety measures in the AHRQ *National Healthcare Quality Report*, MDS measures for long-term care, Joint Commission sentinel events, and the MedMARx Reporting System for Medication Errors.

We consider these patient safety measures to be the best ones currently available for assessing the baseline status of patient safety in the United States and for tracking trends into the future. The measures inevitably involve some limitations in data or methods. Trends in outcomes measures also can be the result of a variety of influences, such as changes in patient safety practices that affect that outcome or changes in reporting practices (without any practice improvement). We have considered such issues in our interpretation of the observed values and trends for the outcomes measures being examined.

Patient Outcome Measures Published by Other Organizations

Patient Safety Outcomes in the *National Healthcare Quality Report*. In 2003, AHRQ began annual publication of the NHQR, a resource that provides trending information on outcomes measures for multiple aspects of health care quality in the United States, including patient safety. The NHQR draws on several data resources to provide summary statistics on patient safety processes and outcomes, including data from the HCUP NIS dataset to measure the PSIs, hospital-acquired infection rates from the National Nosocomial Infections Surveillance System (NNIS), measures of unsafe medication practices from the Medical Expenditure Panel Survey (MEPS), and three measures from the Medicare Patient Safety Monitoring System based on data abstracted from hospital charts. The most recent NHQR, published in 2004, provides national outcomes measures for the years 2001 and 2002. AHRQ also makes available by Internet a set of supplementary NHQR tables with national trend data for many of the patient safety measures.

The NHQR approach of providing snapshots for patient safety outcomes provides readily consumable information for policymakers, but it has the disadvantage of not communicating the sampling, data, and methodological limitations that NHQR has considered in interpreting its findings. For example, the NNIS dataset on nosocomial infections is based on voluntary reporting by a sample of approximately 300 hospitals that are not selected to represent the nation as a whole (Richards et al., 2001). Similarly, the MPSMS identifies patient safety events from chart review for a sample of Medicare patients. This group is very important in its own right, but is not representative of the general American population.

MDS Measures for Long-Term Care. CMS requires all long-term care facilities certified by Medicare or Medicaid to report health-related data on their residents, using the MDS form and data elements. The facilities report MDS data for each resident on at least a quarterly basis. CMS generates annual reports summarizing quarterly trends in the MDS measures nationally and by state, which are regularly published on the CMS Web site.

Two of the measures reported in the MDS are relevant to patient safety: the prevalence of falls and the incidence of pressure ulcers. Both measures reflect injuries that residents may experience as a result of inadequate vigilance in nursing-facility care. However, they offer limited perspective on patient safety outcomes, highlighting the need for development of additional measures and data sources for the long-term care setting. The potential for capturing data for additional patient safety measures from the MDS reporting system merits exploration, including assessment of technical issues affecting data validity and reliability.

To assess baseline experience for falls and pressure ulcers, we abstracted rates based on MDS data published on the CMS Web site. On average, approximately 15 percent of nursing-facility residents in the country were reported to have falls in each quarter during 2000, and the percentage decreased slightly to approximately 13 percent for quarters during 2001 through the first half of 2004. This difference is so small that it may not be a practically significant change, although it is statistically significant due to the very large number of records in the database. An estimated 10 percent of nursing-facility residents nationwide were reported to have experienced pressure ulcers in the period 2000 through 2004, with no observable change over time.

Joint Commission Sentinel Events. The Joint Commission has a long-standing policy for voluntary reporting of sentinel events by the organizations it accredits. The Joint Commission collects information on reported sentinel events in its own database, and it has published summary statistics and trends for the period from 1995 through 2004. Underreporting of events by providers, which is a limitation for any reporting system, means that estimates from the sentinel-events database are underestimates of actual event frequencies. In addition, because the reporting facilities establish their own definitions for sentinel events, there is likely to be some inconsistency in how events are classified. Despite these limitations, the Joint Commission data on sentinel events is a key resource for tracking national trends in reported occurrences for the most serious adverse events.

As of December 2004, the Joint Commission had reviewed a total of 2,966 sentinel events. Of these, more than 75 percent occurred in general hospital or psychiatric hospital settings, and about 75 percent of the events resulted in the death of a patient. About 66 percent of the events were identified through voluntary reports from facilities. Annual reporting rates for sentinel events have climbed substantially during the past decade, a trend that probably reflects some combination of improvement in reporting mechanisms and in patient safety awareness.

Figure 7.2 illustrates the counts for the four types of events with the highest frequencies, which collectively comprise about 50 percent of all events reported. For each category, the frequency of reported incidents increased at varying rates from 1995 through 2004. The Joint Commission data from root cause analyses on sentinel events, reported in Figure 7.3, offer some insights into the relative importance of various factors that contribute to events. Across all types of events, the most frequently identified factors are problems in communication and in staff orientation and training. However, the patterns of root causes differ across event types.

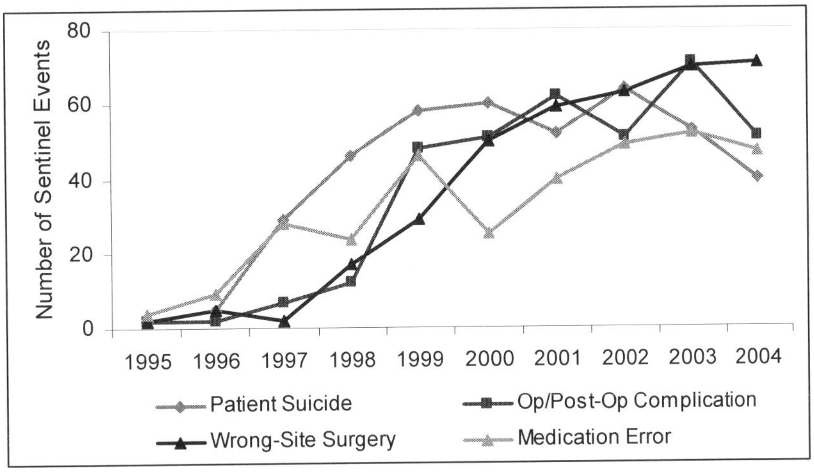

Figure 7.2 Number of Sentinel Events Reported to the Joint Commission, for Top-Four Types of Events, 1995–2004

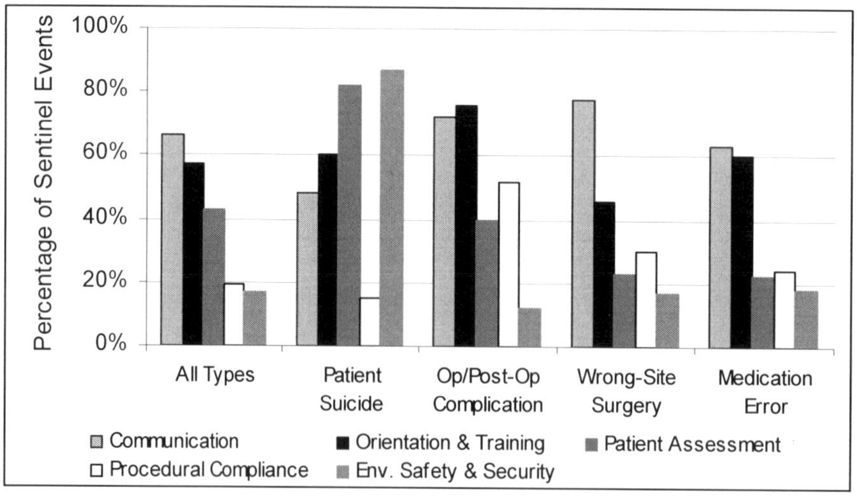

Figure 7.3 Key Root Causes of Sentinel Events Reported to the Joint Commission

MedMARx Reporting System for Medication Errors. MedMARx is a voluntary reporting system for adverse medication events operated by the United States Pharmacopia through its Center for the Advancement of Patient Safety. The MedMARx reporting system originally became available in August 1998, and the number of participating facilities has increased steadily from 56 in 1999 to 570 in 2003. Because reporting is voluntary, however, the participating facilities are not a representative sample of health care institutions in the United States. Medication events captured by MedMARx are classified according to the Index for Categorizing Medication Errors, a taxonomy developed by the National Coordinating Council for Medication Error Reporting and Prevention, which categorizes medication events on a scale ranging from errors that do not reach patients to those that cause varying levels of patient harm.

Figure 7.4 shows that the number of errors reported per facility has grown fourfold since 1999, and the largest rate of growth has been for errors with no harm. From 1999 to 2003, there was a decrease in the share of reported errors that caused harm to patients, from 2.8 percent to 1.5 percent of the totals; there was also a decline in the share of no-harm errors, from 89.2 percent to 82.8 percent. At the same time, the share for events that did not involve any errors doubled from 8.0 percent to 15.7 percent of the total reported events (the events reported are those that potentially or actually cause patient harm).

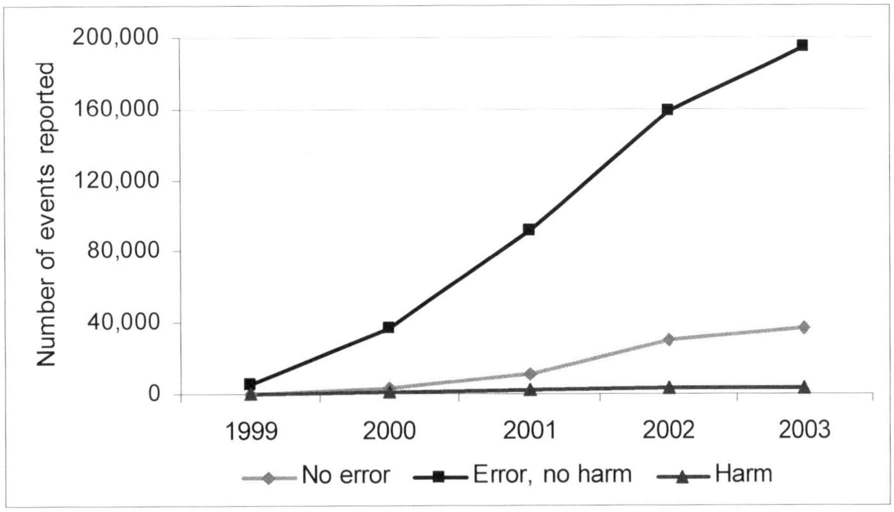

Figure 7.4 Frequency of Medication Events Reported to MedMARx, by Type of Event

Measures of Outcome Rates Estimated in This Evaluation

Two sets of patient safety outcome measures were used in the analysis of baseline trends for measures based on data from inpatient claims or discharge records. The first source was AHRQ's patient safety indicators (PSIs), which were developed using hospital inpatient discharge data from the HCUP NIS. Many aspects of inpatient health care are not addressed by the PSIs, however. Therefore, we identified measures from the second source, which was the set of claims-based outcome measures developed by reporting demonstration grants in Utah and Missouri (UT-MO), with AHRQ patient safety funding. Records for hospital inpatient stays that use the Uniform Billing format are consistent in content and format, enabling aggregation over populations and geographic areas.

Selection of Measures. A key aspect of this analytic work was the selection of a limited number of measures from each measure set for use in our analysis of trends in patient safety outcomes. The goal was to identify subsets of the PSIs and UT-MO measures that most strongly represent patient safety issues. As part of this work, we reviewed several empirical validation studies for the UT-MO measures, and we revised several of the measures to remove codes for events that, by definition, excluded medical error. The measure subsets were selected using information obtained from a review of the existing literature, as well as judgments of importance and clinical relevance by several physicians who are researchers at RAND and experienced in patient safety. Table 7.1 lists the measures identified and used in our trend analysis, along with the at-risk population in 2000 for each measure.

Table 7.1
Selected Patient-Safety-Outcome Measures Used in the Baseline Analysis

Outcome Measure
PSI Measure
Postoperative Hemorrhage or Hematoma
Death in Low-Mortality Diagnosis-Related Groups (DRG)
Postoperative Hip Fracture (not significant)
Postoperative Pulmonary Embolism (PE) or Deep Vein Thrombosis (DVT)
Failure to Rescue
UT-MO Measure
Accidental Cut During Procedure
Other Complications of Procedures
Other Misadventures of Surgical and Medical Care
Respiratory Arrest
Poisonings by Medication
Complications Affecting Specific Body Systems
Alterations in Mental Status
Accidental Falls

These measures are rates, with the denominators differing to reflect the patient populations that are at risk for each type of outcome. The denominators for the PSIs are specified "at risk" populations that are fully defined in the technical specifications for the measures. For the UT-MO measures, the denominators for the two measures related to surgery are the populations of persons who receive surgical care, except those for whom the code that defines the event for a measure is in the primary diagnosis field. That is, the patient already experienced the event (e.g., medication poisoning) before admission. For all the other UT-MO measures, the denominators are the entire populations of persons receiving hospital inpatient services, with the same exclusion for primary diagnostic codes.

Trends for the Selected Rate Measures. We analyzed outcome trends over the period from 1994 through 2002 for the five selected PSI measures and eight selected UT-MO measures, which we calculated using data from HCUP NIS. This approach extends the analysis published by AHRQ in the 2004 NHQR, which reported results for only four time points (1994, 1997, 2000, and 2001). Filling in the entire time series is important for examining trends, to ensure that we can observe any variations in rates from year to year, assess their implications for patient safety trends, and estimate deviations from baseline trends for later years.

Our findings, reported in Table 7.2, show annual percentage changes in occurrence rates for the measures over the nine-year period, with rates increasing for some measures and decreasing for others. In general, the five PSIs show relatively flat adjusted occurrence rates over the 1994–2002 interval. Three of the five PSI measures had small but significant downward trends in their rates of occurrence. Only the "post-operative PE or DVT" had a small but significant upward trend.

Several of the UT-MO measures appear to have had stronger upward or downward trends than the PSIs. In particular, "alterations in mental status," and "other complications of procedures" rates had the largest percentage increase over the nine-year interval, whereas the "respiratory arrest" and "complications affecting specific bodily systems" rates had the largest percentage decrease. These trends could reflect changes over time in actual patient safety outcomes, coding practices, or even International Classification of Diseases, Ninth Revision (ICD-9) coding definitions. Nevertheless, the findings are suggestive of patterns of national improvement in some patient safety measures and deterioration in others for the baseline period.

Table 7.2

Trends in Selected Patient Safety Outcome Measures, 1994–2002

Outcome Measure	Percentage Annual Change
PSI Measures	
Postoperative Hemorrhage or Hematoma	–3.13%
Death in Low Mortality DRGs	–1.47
Postoperative Hip Fracture (not significant)	–1.37
Postoperative PE or DVT	2.29
Failure to Rescue	–2.34
UT-MO Measures	
Accidental Cut during Procedure	1.29
Other Misadventures of Surgical and Medical Care	2.81
Respiratory Arrest	–5.31
Poisonings by Medication	–2.64
Complications Affecting Specific Body Systems	–4.80
Other Complications of Procedures	12.23
Alternations in Mental Status	4.46
Accidental Falls	2.91

NOTE: All annual changes are significant at the 0.01 level except Postoperative Hip Fracture, which is not significant at either the 0.01 or 0.05 level.

The measures also vary widely in frequency of occurrence, with some measures showing far higher rates than others. Two of the PSI measures, "failure to rescue" and "post-operative pulmonary embolism or deep vein thrombosis," occur at far greater (population-adjusted) rates than do the other three PSI measures we studied. Similarly, some of the UT-MO measures (e.g., accidental falls) show far more frequent occurrence rates than do others (e.g., accidental cuts, punctures, perforations, or hemorrhages). These differences in rates are reflected in the different scales used for trends in rates presented in Figures 7.5 and 7.6 (the PSIs) and Figures 7.7 and 7.8 (the UT-MO measures), respectively.

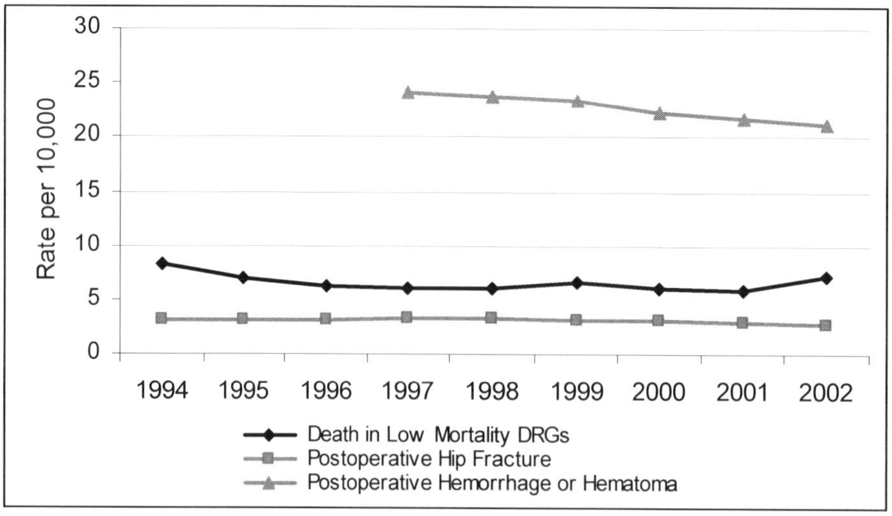

NOTE: According to AHRQ, the measure Postoperative Hemorrhage or Hematoma is not computable for the years 1994 and 1995, because the required diagnosis codes (998.11 and 998.12) became effective in ICD-9 only after 1996.

Figure 7.5 Trends in Selected PSI Measures, 1994–2002 (1)

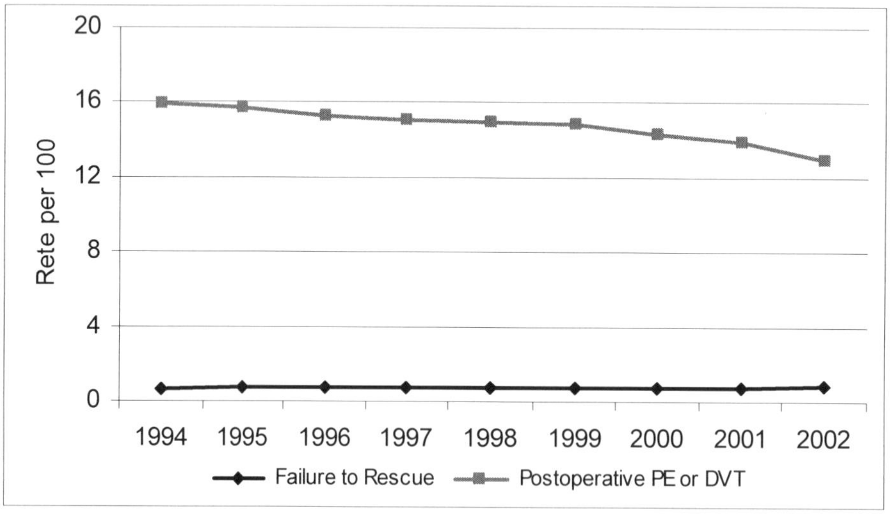

Figure 7.6 Trends in Selected PSI Measures, 1994–2002 (2)

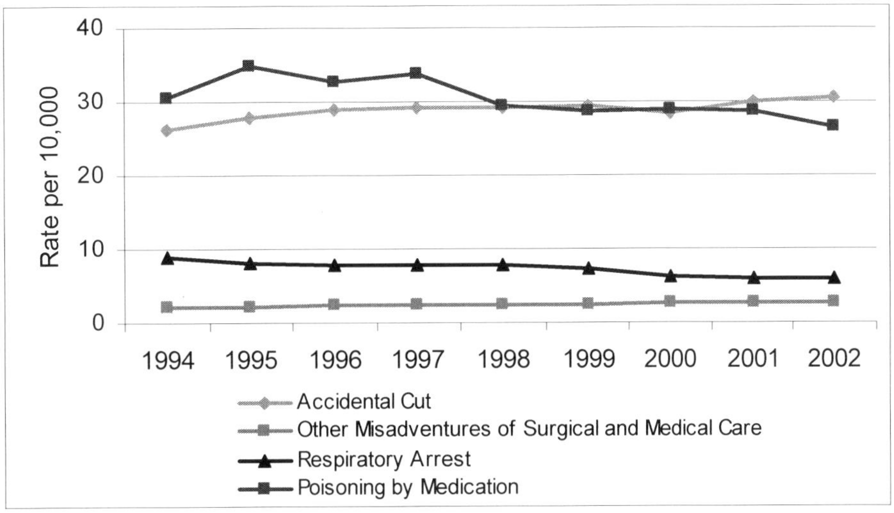

NOTE: Postoperative PE or DVT stands for Postoperative Pulmonary Embolus (PE) or Deep Vein Thrombosis (DVT).

Figure 7.7 Trends in Selected UT-MO Measures, 1994–2002 (1)

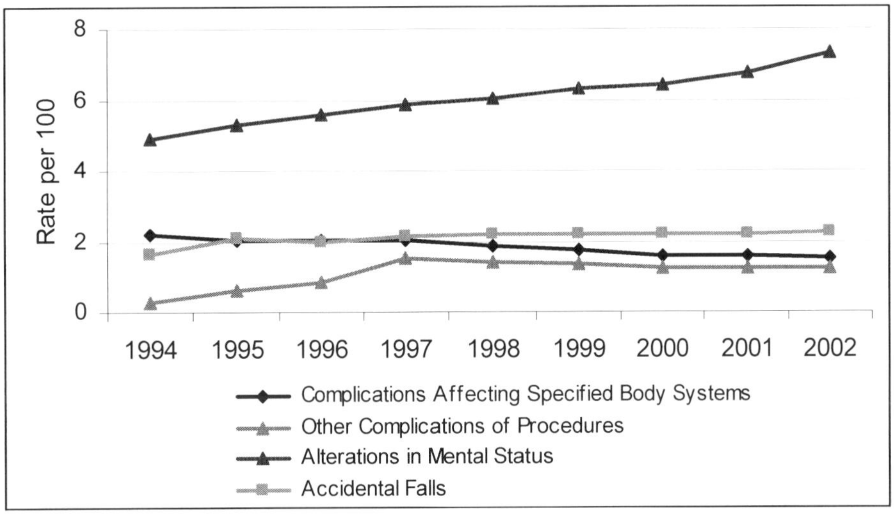

Figure 7.8 Trends in Selected UT-MO Measures, 1994–2002 (2)

Sensitivity to Coding Structure. We also attempted to estimate our selected set of measures using the dataset from the Medicare Current Beneficiary Survey (MCBS), which includes Medicare claims data for a sample of approximately 12,000 beneficiaries. Our goal was to explore the technical challenges of applying the measures to a dataset other than HCUP and to examine whether structural differences between datasets are likely to affect patterns of results on patient safety outcome measures. The HCUP dataset includes 30 data elements containing data for diagnostic and procedural codes from the claims; the MCBS includes only four such codes. Because encounter-based patient safety measures use these data elements to identify adverse events, we expected that a dataset with more fields would produce higher frequencies for observed events.

We also compared estimates from HCUP and MCBS data for the patient safety measures for the elderly population, as well as estimates from HCUP data for the general and elderly populations. This comparison was done as a first step in exploring how the Medicare databases might be used to estimate patient safety measures for ambulatory care, skilled nursing care, and other settings. To the extent that the HCUP and MCBS produce similar patterns of patient safety results from hospital data, we might conclude that future analyses drawing on Medicare *ambulatory* data could offer grounds for inference about ambulatory patient safety outcomes in the general population.

The current analysis of HCUP and MCBS data for the year 2000 showed substantial differences in observed rates of adverse patient safety events, in both comparisons between elderly and general-population samples and comparisons between elderly samples. These results for seven of the outcome measures are shown in Figure 7.9. Similar results were obtained for the remaining six measures, which are not reported here. Outcomes in the HCUP dataset were worse for the elderly than for the general population on all measures studied, with the notable exception of "poisoning by medication," which was twice as frequent in the general population as among the elderly.

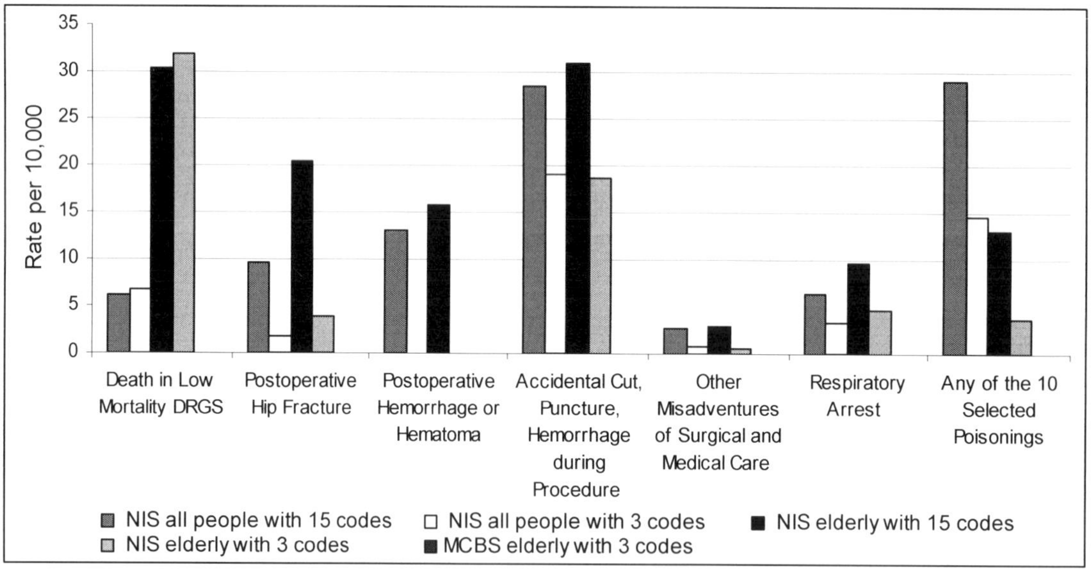

NOTE: The name for the measure "accidental cut, puncture, perforation, or hemorrhage during procedure" has been shortened in the graph to better fit the space available.

Figure 7.9 Estimates of Selected Patient Safety Measures Using Different Samples and Diagnosis Codes

By contrast, our analyses of patient safety measures in the MCBS dataset produced the anomalous finding of zero or near-zero occurrence rates on all of the measures studied. These results affirm the importance of having a full set of diagnostic and procedure codes in the datasets to ensure that the outcome measures are being estimated correctly. They also suggest that it is difficult to identify rare patient safety events using the MCBS because the dataset has a comparatively small sample size (only about 1/500th the size of the HCUP NIS).

To explore further the effects of database structure on patient safety outcomes measures, we compared our findings on the full HCUP dataset, including all 30 available fields for diagnostic and procedural information in healthcare claims data, with a truncated version of the same dataset that included only the first four fields. The results of this analysis also are shown in Figure 7.9. Truncation of the number of available HCUP diagnostic codes generally resulted in lower rates of observed patient safety events. However, the rate of "death in low-mortality DRGs" *increased* with the reduced number of data fields, most likely because many clinical conditions are excluded from the at-risk population for "death in low-mortality DRGs" (e.g., cases involving trauma, immunocompromised states, and cancer). Therefore, truncation of the diagnostic codes reduced the number of observations associated with excludable DRG codes, thus increasing the denominator as well as the numerator for the measure.

Use of Baseline Trends in Rates for Future Outcome Analyses. The analyses of baseline trends in rates for the patient outcome measures reported here served three purposes, two of which were methodological. First, these data provide information for health care providers, policymakers, standard-setting organizations, and other stakeholders involved in patient safety activities regarding historical performance experiences for selected safety outcome measures. Second, the process of calculating the trend data revealed a number of measurement and methodological issues that needed to be addressed to ensure the validity of the trend data used to assess effects. Finally, the historical data is the basis for estimating linear baseline trend lines to which future values for the measures can be compared. To the extent that future values for a measure diverge from its historical trend line, and the differences are statistically significant, we may be observing an effect of patient-safety-improvement activities on that outcome. As data for years following the baseline period start to become available, it will be possible to pursue these analyses.

LESSONS FROM THE BASELINE TREND DATA

Several highlights emerged from the analyses of trends in reported events and rate measures, which will be taken into consideration as the product evaluation moves forward. Perhaps the most obvious one is the large differences in the trends observed for the reported events and measures that are rates. Occurrences of reported events have increased over time for both the Joint Commission sentinel events and the MedMARx adverse medication events. By contrast, trends in rate measures generally have changed more gradually, some moving up and others moving down, while rates for some measures have not changed at all.

The following issues were identified specifically for the *reported event measures*:

- Incomplete reporting by health care providers creates a downward bias in occurrences, which is a serious limitation for using reported-event data to monitor trends in outcomes.
- As the number of providers reporting events increased over time, which occurred for both the Joint Commission and MedMARx data, the total counts escalated simply because of the increased number of reporting organizations, independently of the rate at which each provider reported events.
- Frequencies of reported events also may reflect results of improved patient safety vigilance and reporting practices, which often lead to growth in the number of reported events; this is a positive change for safety improvement, because the events must be identified before providers can act on them. Trends should start to decline eventually as improved practices reduce event occurrences.

The following issues were identified specifically for the *rate measures*:

- Differing trends observed for the PSI and UT-MO rate measures suggest that safety performance varies across measures, and patient safety activities may have differing effects on individual outcomes. Thus, measures used to monitor changes in outcome rates need to be selected carefully to ensure that, as a group, they fulfill the function of serving as proxies for larger performance trends across a health care setting.
- Variations over time in the trends for many of the rate measures highlight the importance of using data for multiple years, to establish valid estimates of baseline trend lines for the rate measures.
- Use of data from different data sources and for multiple years brings with it the challenge of adjusting estimates for changes in definitions for diagnosis codes, as well as in methods for calculating the measures using them.
- For effective interpretation of the baseline trends, and any changes from them, it may be necessary to explore observed trends through both engagement with health care providers and further analyses of the data.

ISSUES AND ACTION OPPORTUNITIES

Analysis of baseline trends is a necessary first step in exploring the effect of AHRQ's patient safety activities, and we will extend the trend data next year. Meanwhile, exploratory analysis of patient-safety outcomes measures based on encounter data helps to assess the available infrastructure for doing national performance monitoring. This kind of baseline analysis also highlights areas for which measures and data infrastructure are currently lacking or for which more development effort is needed in order to make broad-based patient safety measurement possible.

Issues to Consider

Our examination of baseline trends in patient safety outcomes identified some important opportunities for future action by AHRQ. Ambulatory care settings are perhaps the highest priority for development work, but even currently existing hospital encounter-based measures and datasets can benefit from additional refinement, expansion, and validation. In year four, we will expand on the work presented here by exploring the initial effect of AHRQ's activities on patient safety outcomes. We also will explore additional possible sources of patient-safety process and outcomes data, and will make further efforts to contribute to the development of patient safety measures across a range of health care settings and procedures.

Suggestions for AHRQ Action

- **AHRQ should harmonize and validate the capture of claims information in existing inpatient claims databases (in collaboration with CMS and other organizations), by evaluating the consistency of claims-coding practices across hospitals and regions, and by adding to the MCBS claims files the data needed to estimate patient safety measures.**

Experts agree in general that some measures are vulnerable to bias due to variations in coding. However, limited work has been done to quantify this issue for specific measures, which would require assessing the extent of consistency between frequency of events identified in claims data to source information documented in medical records. The MCBS claims dataset is a national-level data resource with potential for use with patient safety measures in several health care settings, especially those targeting the elderly population. However, the data extracted from the source Medicare claims files to populate the MCBS records include only a small subset of the procedure and diagnosis codes available in the source claims. They also exclude variables on admission source and discharge destination, which are available in the source claims and which are required to construct some of the PSIs and other measures, and to adequately measure the incidents of events addressed by these measures.

- **AHRQ should place a priority on developing a set of patient safety measures for ambulatory care settings, and it should foster the establishment of a data infrastructure that can support measurement for ambulatory-care patient-safety issues.**

Only a few measures and very limited data are available for ambulatory-care patient safety, and available measures focus on narrow sets of clinical conditions or processes in care. AHRQ's funding of data-development work by the PBRN coordinating center should begin to address this issue. Any effort to develop such measures will confront the challenge that patient safety in ambulatory care involves interfaces between different clinical specialties or services.

- **AHRQ should work collaboratively to establish an infrastructure and procedures for regular collection of data on the use of effective patient-safety tools and practices by health care organizations, for inclusion in the national network of databases, along with reports from the organizations about the effects of those tools and practices on care processes and clinical outcomes.**

Recognizing that current patient safety activities are in the early phase of what will be continued diffusion over the next several years, RAND has been evaluating alternative approaches for collecting baseline information on these activities as part of the product evaluation. This measurement process should be designed so that AHRQ and collaborating organizations can sustain the data collection after the formal evaluation contract has ended, thereby enabling policymakers and stakeholders to track progress in the adoption of safer practices over time.

CHAPTER 8. CONCLUSION

The evaluation results presented in *Evaluation Report III* have focused on the process and product evaluations in 2004–2005. Some highlights from this phase of the evaluation are assessments of the potential contributions of the health IT projects to patient safety knowledge and practices and AHRQ's progress in activities to disseminate proven patient safety practices for broad adoption by health care providers. We also developed baseline trends for selected measures to assess effects of the initiative on patient outcomes and other stakeholders. In addition, the activities of field-based initiatives (e.g., 5 Million Lives Campaign; see Chapter 6) have become subjects for the evaluation, because they are important vehicles for the diffusion of safe practices across the health care system, and AHRQ is participating in them as a partner as part of its dissemination strategy.

The progress of AHRQ's patient safety initiative in the five years since the publication of the IOM report *To Err Is Human* can be summarized with respect to each of the five system components upon which we organized our process evaluation. Its greatest progress has been in the contributions made to development of knowledge of patient safety epidemiology and effective practices, through the patient safety projects it has funded each year, as well as to development or strengthening of infrastructure to support adoption of safety practices. The components for which progress has been slower are the establishment of a monitoring and vigilance capability and the dissemination of knowledge and products into the field for use by health care providers and other end users.

As the patient safety initiative proceeded, AHRQ has made a shift in its activity mix, with stronger emphasis on the synthesis of knowledge and products emerging from its funded projects, along with packaging and dissemination of products and tools for adoption by health care providers. Results from the FY 2000–FY 2001 projects are entering the published literature rapidly, which will need to be synthesized to provide the information base for the dissemination and adoption strategies. Other AHRQ activities—such as the PSIC and the hospital culture survey—have come to maturity and are beginning to build infrastructure and influence patient safety practices across the country. At the same time, ongoing grant projects are addressing health IT and implementing patient safety practices, continuing to build knowledge and evidence to support practice improvements.

Given the size of the U.S. health care system and the patient safety problems it has been documented to have, it is a daunting task to make the system safer. With small funding relative to the size of the problem, the potential for AHRQ to have an effect on creating a safer health care system may be limited. Recognizing its funding constraints, AHRQ has been using a variety of approaches to leverage resources, such as cost sharing on implementation projects and working through partners.

Realistically, the achievement of the goal of a safer health care system will depend on the commitment of thousands of organizations around the country, with leadership and technical support from agencies such as AHRQ and other federal and state agencies. The momentum of widespread commitment and participation in field-based initiatives has become fully visible this year, which bodes well for diffusion of safer practices. The enactment of PSQIA has created another important mechanism to make it safer for providers to report safety events and to work to prevent them. It is hoped that this mechanism also will provide a much-needed stimulus for establishment of a national patient safety data repository, which we believe is essential to ensuring the sustainability of effective patient-safety vigilance and protections over time.

FUTURE DIRECTIONS AND PRIORITIES

From our assessments of AHRQ's patient safety strategy, as well as the current activities of its grantees and field organizations, we have identified several priorities that ARHQ is encouraged to pursue in the near future:

- Facilitate movement toward a *national network of patient safety databases* by using the provisions in the Patient Safety and Quality Improvement Act of 2005 to encourage use of consistent data standards, and establish a set of national patient safety measures for assessing performance.
- *Identify key patient-safety practices and products* from the results of the FY 2000–FY 2001 patient safety projects by synthesizing results from groups of projects addressing similar issues or practices.
- *Package and disseminate* patient safety products that derive from the synthesis of project results, including development of "off-the-shelf" products that can be used readily by health care organizations.
- Update the *patient safety evidence report* to incorporate recently published results from the patient safety projects, applying standards of evidence that ensure rigorous assessment of study designs for testing patient safety practices that cannot be tested effectively using randomized control study designs.
- Assess the role of *health information technology* in achieving safer health care practices and its interface with the human aspects of care delivery, using results of the newly funded health IT grants, as well as knowledge generated by other patient safety projects that have addressed use of technology for patient safety practices.
- Continue to *engage actively in field-based partnerships* that enable ARHQ to optimize its contribution to stimulating broad adoption of proven patient safety practices by health care providers, within the constraints of its finite resources.

NEXT STEPS FOR THE EVALUATION

In 2005–2006, the patient safety evaluation center will perform the last year of the four-year evaluation. We will continue gathering information on the evolution of the patient safety initiative, and we also will consolidate results across the four years of the evaluation to present cumulative information on the progress and effects of the patient safety initiative. Our work on the product evaluation will continue, moving the impact assessment as far as possible in this limited time frame. We will suggest structures and processes that AHRQ can use to continue monitoring effects of the initiative on stakeholders and patient outcomes.

Evaluation questions we will address encompass both the process and product phases of the patient safety evaluation:

- How is the scientific-evidence base being expanded by new knowledge on patient safety epidemiology and practices from the AHRQ-funded patient safety projects?
- What progress is being made in building a national patient safety infrastructure?

- What actions is AHRQ taking with other organizations to disseminate knowledge, develop products for improving patient safety practices, and provide training for users to encourage broad adoption of proven patient safety practices?
- What effects are the collective activities of AHRQ, other federal and state agencies, and health care organizations across the country having on stakeholders, including patient outcomes and effects on other stakeholders?

For the process evaluation, we will synthesize the four-year progress with respect to each component of our framework for an effective patient safety system. We also will attempt to assess the experiences and lessons from the field-based patient safety initiatives discussed for the first time in Chapter 6 of this report, which we view as being central to successful diffusion of safe practices across providers. We will do so in tandem with continued attention to AHRQ's strategy and activities in disseminating results from its funded projects; activities and impacts of the PSIC; and developments from its funded grants.

APPENDIX
SUGGESTIONS FOR AHRQ ACTION: *EVALUATION REPORTS I* AND *II*

Monitoring and Vigilance	
***Evaluation Report I*: Sept 2003**	***Evaluation Report II*: Sept 2004**
AHRQ should actively facilitate development work toward establishment of a national patient safety data repository, using a structured consensus process to select national measures, develop the data standards and specifications, and delineate procedures for operation of the reporting network. In identifying candidate measures, AHRQ should ensure that the most important safety aspects of the patient's health care experience are identified and represented by the measures. The process used by University of California at San Francisco (UCSF) and Stanford University to select the Healthcare Cost and Utilization Project (HCUP) patient safety indicators should be adapted by AHRQ for selection and definition of the broader set of national patient safety measures. AHRQ should place a priority on the synthesis it plans to prepare of the knowledge, products, and newly tested reporting systems that will begin to emerge from the Reporting Demonstration projects during FY 2004. AHRQ should invite accreditation and credentialing organizations and insurers to be actively involved in the process for establishing national patient safety measures and designing a reporting network, with the goal of their adopting the measures as standards in their accreditation processes.	As the state and regional health-information-systems projects progress, AHRQ should leverage this work to encourage broad use of the data standards recommended in the IOM report *Patient Safety: Achieving a New Standard of Care*. AHRQ should build on the technical products of the federal data system project by pursuing expanded use of the newly developed reporting and data warehouse capability, with the goal of moving toward a national data repository with multiple public and private users. Using a structured consensus process involving multiple stakeholders, AHRQ should place a priority on building on the existing Patient Safety Indicators to establish a broader set of national patient safety measures that represent the most important safety aspects of the patient's health care experience in a variety of settings. AHRQ should invite accreditation and credentialing organizations and insurers to be actively involved in the process for establishing national patient safety indicators and designing standards for national data reporting, with the goal of their adopting the indicators and standards in their accreditation processes.

Patient Safety Epidemiology	
***Evaluation Report I*: Sept 2003**	***Evaluation Report II*: Sept 2004**
AHRQ should ensure that the results of epidemiological studies by the patient safety projects are summarized in usable forms for a variety of stakeholders and for future decisions on patient safety priorities.	AHRQ should ensure that the results of epidemiological studies by the patient safety projects are summarized in usable forms for a variety of stakeholders and for future decisions on patient safety priorities.

AHRQ should establish definitions and standards for measurement methods as the bases for valid and consistent epidemiologic estimates for patient safety issues. AHRQ should fund the development of a review report that summarizes the current state of knowledge on patient safety epidemiology and presents the best available estimates of the incidence and severity of errors and adverse events. Future AHRQ funding for research on patient safety epidemiology should focus on topics that have been addressed least frequently thus far, placing a priority on areas that have high risk for patient harm.	AHRQ should establish definitions and standards for measurement methods as the bases for valid and consistent epidemiological estimates for patient safety issues. AHRQ should fund the development of a review report that summarizes the current state of knowledge on patient safety epidemiology and presents the best available estimates of the incidence and severity of errors and adverse events.
Effective Practices and Tools	
***Evaluation Report I*: Sept 2003**	***Evaluation Report II*: Sept 2004**
As the AHRQ-funded patient safety projects and other research generate new evidence on the effective practices, AHRQ should update the evidence report on patient safety to incorporate this evidence and should make the evidence information readily available to users through Web-based and other communication media. AHRQ should commit resources to defining the standards of evidence that should apply for assessing the effectiveness of patient safety practices, and to determining methods for prioritizing which practices should be implemented. Future AHRQ research funding should focus on testing patient safety practices that are promising, but for which evidence regarding their effects on safety outcomes is lacking or insufficient. AHRQ should support work to document the costs, cost-effectiveness, and return on investment of promising patient safety practices, to make the business case for their adoption in the field. To support health care organizations in adopting evidence-based patient safety practices, AHRQ should collaborate with users to establish implementation guides and tools with practical "how to" information that reduces barriers to implementation progress.	AHRQ should commit resources to defining the standards of evidence that should apply for assessing the effectiveness of patient safety practices. To this end, AHRQ should support a panel process to produce recommendations regarding standards of evidence for patient safety. As the patient safety projects generate new evidence on practices and standards of evidence have been adjusted to apply more effectively to patient safety practices, AHRQ should update the evidence report on patient safety to incorporate new evidence for widespread availability to users. AHRQ should pursue a twofold strategy to generate information on the business case for promising patient safety practices: (1) Require all of its funded patient safety projects that are conducting practice interventions to collect and report data on implementation costs as part of their research; and (2) identify some of the projects that have successful interventions and separately fund analyses of the cost-effectiveness and return on investment for those interventions. For subsequent patient safety implementation grants, AHRQ should focus on funding efforts by nonacademic medical centers in order to improve the generalizability of findings on patient safety practices.

Building Infrastructure for Effective Practices	
***Evaluation Report I*: Sept 2003**	***Evaluation Report II*: Sept 2004**
AHRQ should undertake a variety of partnerships with public- and private-sector organizations to explore creative ways for putting tested new patient-safety products and knowledge into practice in the health care system. AHRQ should pursue a focused strategy to integrate the patient perspective into activities to improve patient safety performance, including relevant research, reporting by consumers on their experiences with care, and consumer involvement in activities by health care organizations. AHRQ should explore mechanisms for establishing contingency financing of dissemination and implementation activities to provide flexibility to respond to needs or opportunities identified in the field. Continued funding support should be provided for Developing Center grants and other projects that are beginning to build patient safety research infrastructure, to enable them to become self-sustaining. AHRQ should expand its efforts to take full advantage of its already-established network infrastructures to serve as testing grounds for new patient safety interventions.	AHRQ should strategically seek out new partnerships, especially in areas in which little collaboration currently exists, while strengthening existing partnerships. Wherever possible, AHRQ should eliminate real and perceived barriers, by other organizations (private or public), to partnering with them. AHRQ should seek ways to maintain and build on the network of trainees who have gone through the Patient Safety Improvement Corps training. AHRQ should expand the Patient Safety Improvement Corps model to include stakeholders in addition to state governments and hospitals. AHRQ should fund Centers of Excellence for Consumer Engagement to study the effect of involving patients and families in patient safety activities. AHRQ should partner with consumer organizations and organizations with expertise involving patients and families to disseminate best practices for consumer engagement in patient safety improvement. AHRQ should encourage the use and evaluation of information technology to increase consumer awareness of patient safety issues and provide a means for consumers to report errors at the time they occur.

Broader Adoption of Effective Practices	
***Evaluation Report I*: Sept 2003**	***Evaluation Report II*: Sept 2004**
AHRQ should design and carry out a cohesive strategy to disseminate the new knowledge and products of the patient safety projects to the broad spectrum of stakeholders that will put them into practice in the delivery of safer health care. AHRQ should identify and test alternative strategies for AHRQ to serve as a catalyst that motivates and supports initiatives by health care organizations to implement new patient safety practices. AHRQ should develop a monitoring process to measure the effects of individual implementation initiatives on patient safety practices and	AHRQ should develop and implement a strategic plan that specifies how the agency will disseminate new patient safety knowledge and products to the broad spectrum of stakeholders, and the actions it will take to facilitate adoption of new, safer practices. AHRQ should expand its internal infrastructure and budget to support the coming knowledge transfer and dissemination work, so that the work is funded appropriately, has effective leadership and appropriate expertise to conduct the work, and has the support of the agency director.

outcomes, as well as the cumulative effects of all the patient safety activities on key outcome measures.	Investment in AHRQ's existing programmatic vehicles should be expanded, strategically using the vehicles to promote the translation of patient safety research into practice, with more-specific guidance on which patient safety applications should be pursued. AHRQ should develop "mentoring grants" that extend the successful work of implementation grantees more broadly across the health care system by enabling them to provide implementation support to other organizations. AHRQ should seek to build partnerships with health-care-system end users to secure their input at the front end of the research process (so that research products are end user–driven) and to help AHRQ with translation and diffusion work by extending the resources and reach of the agency).
Product Evaluation of Initiative Effects	
***Evaluation Report I*: Sept 2003**	***Evaluation Report II*: Sept 2004**
None	AHRQ should develop Consumer Assessment of Healthcare Providers and Systems (CAHPS®) surveys or survey modules for patients to report on patient safety issues in ambulatory care, hospital services, and long-term care settings. AHRQ should work with organizations in the field to initiate measurement capabilities for tracking effects for which data sources do not yet exist.

REFERENCES

Agency for Healthcare Research and Quality, *AHRQ's Patient Safety Initiative: Building Foundations, Reducing Risk*, An Interim Report to the United States Senate Committee on Appropriations, Rockville, Md., 2003b.

Agency for Healthcare Research and Quality, *2004 National Healthcare Disparities Report*, Rockville, Md.: Department of Health and Human Services, AHRQ No. 05-0014, 2004a.

Agency for Healthcare Research and Quality, *2004 National Healthcare Quality Report*, Rockville, Md.: Department of Health and Human Services, AHRQ No. 05-0013, 2004b.

Agency for Healthcare Research and Quality, *Partnerships in Implementing Patient Safety*, Rockville, Md.: Department of Health and Human Services, AHRQ RFA No. RFA-HS-05-012, 2004c.

Agency for Healthcare Research and Quality, *Advances in Patient Safety: From Research to Implementation*, Volumes 1–4, Rockville, Md.: AHRQ Publication Nos. 050021 (1-4), February 2005a.

Agency for Healthcare Research and Quality, *Hospital Survey on Patient Safety Culture*, Rockville, Md., April 2005b. As of August 21, 2007: http://www.ahrq.gov/qual/hospculture/

Agency for Healthcare Research and Quality, *Technical Assistance Briefings on the Hospital Survey on Patient Safety Culture*, Rockville, Md., 2005c. As of September 2007: http://www.ahrq.gov/qual/tacall3ptsaf.htm

Agency for Healthcare Research and Quality, *Developing Evidence to Inform Decisions About Effectiveness: The DEcIDE Network*, Rockville, Md., Request for Proposal No. AHRQ-05-0010, June 2005d.

Agency for Healthcare Research and Quality, *Clinical Decisions and Communications Science Center*, Rockville, Md., Request for Proposal No. AHRQ-05-0013, June 2005e.

Agency for Healthcare Research and Quality, *Accelerating Change and Transformation in Organizations and Networks (ACTION)*, Rockville, Md., Request for Proposal No. AHRQ-06-00001, July 2005f.

Altman, D. E., C. Clancy, and R. J. Blendon, "Improving Patient Safety—Five Years after the IOM Report," *New England Journal of Medicine,* Vol. 351, November 11, 2004, p. 20.

Bessant, J., R. Kaplinsky, and M. Morris, "Developing Capability Through Learning Networks," *Journal of Technology Management and Sustainable Development,* Vol. 2, No. 1, 2003, pp. 19–38.

Bhasale, A. L., G. C. Miller, S. E. Reid, and H. C. Britt, "Analysing Potential Harm in Australian General Practice: An Incident-Monitoring Study," *Medical Journal of Australia*, Vol. 169, No. 2, 1998, pp. 73–76.

Blake, S. C., S. Kohler, K. Rask, A. Davis, and D. V. Naylor, "Facilitators and Barriers to 10 National Quality Forum Safe Practices," *American Journal of Medical Quality,* Vol. 21, No. 5, 2006, pp. 323–334.

Brailer, D. J., "The Decade of Health Information Technology: Delivering Consumer-centric and Information-rich Health Care," Framework for Strategic Action, Office of National Coordinator for Health Information, July 2004. As of September 2007: http://www.hhs.gov/healthit/framework.html#report

Centers for Medicare & Medicaid Services, *Quality Improvement Organizations.* As of September 2005: http://www.cms.hhs.gov/qio/

Chang, A., P. M. Schyve, R. J. Croteau, D. S. O'Leary, and J. M. Loeb, "The JCAHO Patient Safety Event Taxonomy: A Standardized Terminology and Classification Schema for Near Misses and Adverse Events," *International Journal for Quality in Health Care,* Vol. 17, No. 2, 2005, pp. 95–105.

Cooperative State Research, Education, and Extension Service, homepage, U.S. Department of Agriculture. As of August 20, 2007: http://www.csrees.usda.gov/

Dovey, S. M., D. S. Meyers, R. L. Phillips, L. A. Green, G. E. Fryer, J. M. Galliher, et al., "A Preliminary Taxonomy of Medical Errors in Family Practice," *Quality Safety Health Care,* Vol. 11, No. 3, 2002, pp. 233–238.

Encinosa, W. E., and D. M. Bernard, "Hospital Finances and Patient Safety Outcomes," *Inquiry,* Vol. 42, No. 1, 2005, pp. 60–72.

Farley, D. F., S. C. Morton, C. Damberg, A. Fremont, S. Berry, M. Greenberg, et al., *Assessment of the National Patient Safety Initiative: Context and Baseline, Evaluation Report I,* Santa Monica, Calif.: RAND Corporation, TR-203-AHRQ, 2005.

Farley, D. F., S. C. Morton, C. Damberg, S. Ridgely, A. Fremont, M. Greenberg, et al., *Assessment of the National Patient Safety Initiative: Moving from Research to Practice, Evaluation Report II (2003–2004),* Santa Monica, Calif.: RAND Corporation, TR-463-AHRQ, 2007.

Fernald, D. H., W. D. Pace, D. M. Harris, D. R. West, D. S. Main, and J. M. Westfall, "Event Reporting to a Primary Care Patient Safety Reporting System: A Report from the ASIPS Collaborative," *Annotated Family Medicine,* Vol. 2, No. 4, 2004, pp. 327–332.

Galt, K. A., A. M. Rule, B. E. Clark, J. D. Bramble, W. Taylor, and K. G. Moores, "Medication Safety in the Primary Care Physician's Office," paper presented at *Patient Safety 2004: Connecting the Dots to Reduce Harm,* Association for Healthcare Research and Quality, 2005.

Greenhalgh, T., G. Robert, F. Macfarlane, P. Bate, and O. Kyriakidou, "Diffusion of Innovations in Service Organizations: Systematic Review and Recommendations," *Milbank Quarterly,* Vol. 82, No. 4, 2004, pp. 581–629.

Healthcare Information and Management Systems Society, *About HIMSS.* As of August 21, 2007:
http://www.himss.org/ASP/abouthimsshome.asp/

Healthcare Information and Management Systems Society, HIMMS Declaration, *Bridging the Chasm: Realizing a Universal EHR*, March 28, 2003. As of August 21, 2007:
http://www.himss.org/content/files/EHR_Declaration.pdf/

Institute for Healthcare Improvement, *The 5 Million Lives Campaign.* As of September 2005:
http://www.ihi.org/IHI/Programs/Campaign/

Institute of Medicine, Committee on Quality of Health Care in America, *Crossing the Quality Chasm: A New Health System for the 21st Century*, Washington, D.C.: National Academy Press, 2002.

Institute of Medicine, *To Err Is Human: Building a Safer Health System,* L. T. Kohn, J. M. Corrigan, and M. S. Donaldson, eds., Washington, D.C.: National Academy Press, 2000.

Institute of Medicine, *Patient Safety: Achieving a New Standard of Care*, P. Aspden, J. M. Corrigan, J. Wolcott, and S. M. Erickson, eds., Washington, D.C.: National Academy Press, 2004.

Joint Commission on Accreditation of Healthcare Organizations, *Patient Safety Goals.* As of September 2005:
http://www.jcaho.org/accredited+organizations/patient+safety/05+npsg/intro.htm/
As of August 21, 2007:
http://www.jointcommission.org/GeneralPublic/NPSG/

Kizer, K. W., R. Y. Nishimi, E. J. Power, et al., *Standardizing a Patient Safety Taxonomy*, National Quality Forum, Washington D.C.: NQFCR-13-06, 2006.

Lapetina, E. M., and E. M. Armstrong, "Preventing Errors in the Outpatient Setting: A Tale of Three States," *Health Affairs* (Millwood), Vol. 21, No. 4, 2002, pp. 26–39.

Leape, L. L., and D. M. Berwick, "Five Years After *To Err Is Human*: What Have We Learned?" *Journal of the American Medical Association*, Vol. 293, 2005, pp. 2384–2390.

Leape, L. L., G. Rogers, D. Hanna, P. Griswold, F. Federico, C. A. Fenn, et al., "Developing and Implementing New Safe Practices: Voluntary Adoption Through Statewide Collaboratives," *Quality and Safety in Health Care,* Vol. 15, No. 4, 2006, pp. 289–295.

The Lewin Group, Inc., *Health Information Technology Leadership Panel, Final Report*, Falls Church, Va., March 2005.

Longo, D. R., J. E. Hewett, B. Ge, and S. Schubert, "The Long Road to Patient Safety: A Status Report on Patient Safety Systems," *JAMA,* Vol. 294, No. 22, pp. 2858–2865.

McDonald, K., P. Romano, J. Geppert, et al., *Measures of Patient Safety Based on Hospital Administrative Data: The Patient Safety Indicators,* Technical Review 5 (prepared by University of California at San Francisco-Stanford Evidence-based Practice Center, under Contract No. 290-97-0013), Rockville, Md.: Agency for Healthcare Research and Quality, AHRQ Publication No. 02-0038, 2002.

McFadden, K. L., G. N. Stock, C. R. Gowen, "Exploring Strategies for Reducing Hospital Errors," *Journal of Healthcare Management*, Vol. 51, No. 2, 2006, pp. 123–135, discussion p. 136.

Moore, G. C., and I. Benbasat, "Development of an Instrument to Measure the Perceptions of Adopting an Information Technology Innovation," *Information Systems Research,* Vol. 2, No. 3, 1991, pp. 192–222.

Mor, V., "A Comprehensive Clinical Assessment Tool to Inform Policy and Practice: Applications of the Minimum Data Set," *Medical Care*, Vol. 42, No. 4 Suppl, 2004, pp. III50–59.

Mor, V., J. Angelelli, R. Jones, J. Roy, T. Moore, and J. Morris, "Inter-rater Reliability of Nursing Home Quality Indicators in the U.S. BMC," *Health Services Resources*, Vol. 3, No. 1, 2003, p. 20.

Mor, V., K. Berg, J. Angelelli, D. Gifford, J. Morris, and T. Moore, "The Quality of Quality Measurement in U.S. Nursing Homes," *Gerontologist*, Vol. 43, No. 2, 2003, pp. 37–46.

National Quality Forum, *Serious Reportable Events in Healthcare: A Consensus Report*, Washington, D.C., March 2002.

Office of the National Coordinator of Health IT (ONC) Web site. As of September 2007: http://www.hhs.gov/healthit/onc/mission

Quality Interagency Coordination Task Force (QuIC), *Doing What Counts for Patient Safety: Federal Actions to Reduce Medical Errors and Their Impact,* Report of the Quality Interagency Coordination Task Force (QuIC) to the President, 2000.

Richards, C., T. G. Emori, J. Edwards, S. Fridkin, J. Tolson, and R. Gaynes, "Characteristics of Hospitals and Infection Control Professionals Participating in the National Nosocomial Infections Surveillance System 1999," *American Journal of Infectious Control*, Vol. 29, No. 6, 2001, pp. 400–403.

Rogers, E. M., *Diffusion of Innovations,* 5th ed., New York: Free Press, 2003.

Rogers, E. M., J. D. Eveland, and A. S. Bean, *Extending the Agricultural Extension Model: Preliminary Draft*, Stanford, Calif.: Stanford University, Institute for Communication Research, September 1976.

Roy, J., and V. Mor, "The Effect of Provider-Level Ascertainment Bias on Profiling Nursing Homes," *Stat Medicine*, 2005.

Shojania, K. G., B. W. Duncan, K. M. McDonald, and R. M. Wachter, *Making Health Care Safer: A Critical Analysis of Patient Safety Practices*, Evidence Report/Technology Assessment No. 43 (prepared by University of California at San Francisco-Stanford Evidence-based Practice Center, under Contract No 290-97-0013), Rockville, Md.: Agency for Healthcare Research and Quality, AHRQ Publication No. 01-E058, 2001.

Snyder, W. M., and X. de S. Briggs, *Communities of Practice: A New Tool for Government Managers*, IBM Center for the Business of Government, 2003.

Sprague, L., "Contracting for Quality: Medicare's Quality Improvement Organizations," Washington, D.C.: National Health Policy Forum, Issue Brief No. 774, June 2002. As of August 21, 2007:
http://www.nhpf.org/pdfs_ib/IB774_QIOs_6-3-02.pdf/

Stufflebeam, D., W. J. Gephart, E. G. Guba, R. L. Hammond, H. O. Merriman, and M. M. Provus, *Educational Evaluation and Decision Making*, Itasca, Ill.: F. E. Peacock Publishers, Inc., 1971.

Stufflebeam, D. L., G. F. Madaus, T. Kellaghan, *Evaluation Models: Viewpoints on Educational and Human Services Evaluation.* Norwell, Mass.: Kluwer Academic Publishers, 2000.

Surgical Care Improvement Project, *Surgical Care Improvement Project: A National Partnership.* As of August 15, 2002:
http://www.medqic.org/scip/

Tamuz, M., and E. J. Thomas, "Classifying and Interpreting Threats to Patient Safety in Hospitals: Insights from Aviation," *Journal of Organizational Behavior,* Vol. 27, 2006, pp. 919–940.

Valente, T. W., and R. L. Davis, "Accelerating the Diffusion of Innovations Using Opinion Leaders," *The Annals of the American Academy of Political and Social Science*, Vol. 566, November 1999, pp. 55–67.

Wachter, R. M., "The End of the Beginning: Patient Safety Five Years After *To Err Is Human*," *Health Affairs,* Web Exclusive, November 2004, pp. W4-534–545.

Wenger, E., R. McDermott, and W. M. Snyder, "Seven Principles for Cultivating Communities of Practice," *Cultivating Communities of Practice: A Guide to Managing Knowledge*, Cambridge, Mass.: Harvard Business School Press, 2002.